Augsburger Schriften zur Mathematik, Physik und Informatik
Band 31

Edited by:
Professor Dr. B. Schmidt
Professor Dr. B. Aulbach
Professor Dr. F. Pukelsheim
Professor Dr. W. Reif
Professor Dr. D. Vollhardt

Bibliographic information published by the Deutsche Nationalbibliothek

The Deutsche Nationalbibliothek lists this publication in the
Deutsche Nationalbibliografie; detailed bibliographic data are
available in the Internet at http://dnb.d-nb.de .

ISBN 978-3-8325-4363-1
ISSN 1611-4256

Logos Verlag Berlin GmbH
Comeniushof, Gubener Str. 47,
10243 Berlin
Tel.: +49 030 42 85 10 90
Fax: +49 030 42 85 10 92
INTERNET: http://www.logos-verlag.de

# Connecting Atomistic and Continuum Models of Nonlinear Elasticity Theory

## Rigorous Existence and Convergence Results for the Boundary Value Problems

### Dissertation

zur Erlangung des akademischen Grades

### Dr. rer. nat.

eingereicht an der

Mathematisch-Naturwissenschaftlich-Technischen Fakultät
der

## UNIVERSITÄT AUGSBURG

von

### Julian Braun

Augsburg, Juni 2016

Universität Augsburg
Mathematisch-Naturwissenschaftlich-
Technische Fakultät

Gutachter:

- Prof. Dr. Bernd Schmidt, Universität Augsburg

- Prof. Dr. Dirk Blömker, Universität Augsburg

- Prof. Dr. Florian Theil, University of Warwick (Vereinigtes Königreich)

Datum der mündlichen Prüfung: 20. September 2016

# Contents

# Introduction

A material behaves elastically if it deforms under applied external forces but reverts to its original state once these forces are removed. This is often related to the fact that the internal, microscopic structure of the material mostly stays the same even though the material deforms macroscopically. The subject of elasticity theory is to give conditions under which a material behaves elastically and to predict the precise deformation of the material given the applied body forces and imposed boundary conditions.

In classical continuum mechanics, the behavior of an elastic solid is described in terms of a deformation mapping which satisfies the partial differential equations of nonlinear elasticity theory. The stress tensor appearing in these equations depends in a local but nonlinear way on the gradient of the deformation map. For so-called hyperelastic materials, which we will consider in the following, the equations of elastostatics are the Euler-Lagrange equations of an associated energy functional in which the stored elastic energy is given in terms of an integral of an energy density function acting on the local deformation gradient. Stable configurations are given by deformations which are local minimizers of this functional. The stored energy density in particular induces the stress-strain relation and thus encodes the elastic properties of the material. In elastodynamics, the same quantities combined with Newton's second law of motion give rise to the nonlinear evolution equations describing the dynamic behavior of the material. Here, the internal forces are given by the first variation of the elastic energy.

At the same time, a crystalline solid may be viewed microscopically not as a continuum but as a particle system consisting of interacting atoms on a portion of a lattice. The interatomic cohesive and repulsive forces, which are dominantly induced by the atomic electronic structure, can effectively be modeled in terms of classical interaction potentials. Stable configurations are then local minimizers of the total atomistic energy which is given as a sum of all the local interactions. They are also solutions of the high dimensional system of equations for the force balance. The dynamic behavior of the material is again described by Newton's second law of motion where the internal forces are now given by the first variation of the new atomistic interaction energy.

The classical connection between atomistic and continuum models of nonlinear elasticity is provided by the Cauchy-Born rule: The deformations of the single atoms are assumed to follow the gradient of a macroscopic deformation. The continuum stored energy function associated to a macroscopic affine map is then given by the energy (per unit volume) of a crystal which is homogeneously deformed with the same affine mapping. In particular, this entails the assumption that there are no fine scale oscillations on the atomic scale. We will call this the Cauchy-Born energy density in the following. Indeed, if one assumes the Cauchy-Born rule to hold true and consequently requires that the individual atoms follow a smooth macroscopic deformation mapping, one can derive a continuum energy expression for such a deformation from given atomistic potentials as shown by Blanc, Le Bris and Lions, cf. [BLL02]. To leading order in the small lattice spacing parameter $\varepsilon$ one then obtains the continuum energy functional with the Cauchy-Born energy density. However, it is not clear a priori under what conditions the Cauchy-Born hypothesis holds true. Moreover, it is desirable to not only obtain a pointwise convergence result for the corresponding energy functionals, but also to relate the solutions of the continuum problem to those of the atomistic system.

Our aim in this work is to establish rigorous links between the solutions of the atomistic problems in their asymptotic regime $\varepsilon \to 0$ and the corresponding solution of the continuum Cauchy-Born model for the nonlinear elastic behavior of crystalline solids accounting for body forces and boundary values.

Over the last 15 years in particular there has been considerable progress in identifying conditions which allow for a mathematically rigorous justification of the Cauchy-Born rule. Here, we restrict our attention to those contributions which directly have influenced our results. For a more general review on the Cauchy-Born rule we refer to the survey article [Eri08] by Erickson.

In their seminal contribution [FT02] Friesecke and Theil consider a two-dimensional mass-spring model and prove that in their model the Cauchy-Born rule does indeed hold true for small strains, while it fails for large strains. Their result has then been generalized to a wider class of discrete models and arbitrary dimensions by Conti, Dolzmann, Kirchheim and Müller in [CDKM06]. More specifically, the version of the Cauchy-Born rule established in these works is even stronger than what we described above: the authors prove under suitable assumptions that on a box containing a portion of a Bravais lattice under given affine boundary conditions the global minimizer of the elastic energy is given by that same affine deformation, now acting on all atoms.

In the first part of this thesis we establish the precise connection of these results with what we called the Cauchy-Born rule above. More importantly, we will use them in a rigorous convergence result of the atomistic problems to a continuum problem as the interatomic distance $\varepsilon$ goes to zero to prove that, at least for small strains, the continuum energy density is indeed given by the Cauchy-Born rule energy density. A corresponding discrete-to-continuum convergence result in which simultaneously the strain becomes infinitesimally small has previously been proven with mostly quite different methods in [Sch09] resulting in a continuum energy functional with the linearized Cauchy-Born energy density.

Such convergence results are global variational convergence results for the elastic energy in the sense of Γ-convergence. Our Γ-convergence result will be accompanied by a corresponding compactness result, which then automatically ensures the convergence of subsequences of the global atomistic minimizers to a global minimizer of the continuum problem. For a definition of Γ-convergence and more details we refer to Section 1.2 or the literature, e.g. [DM93].

A general approach to prove Γ-convergence results within a class of integral functionals satisfying certain certain growth and boundedness conditions has been developed in depth in the literature, cf. [DM93] or [BD98]. It is sometimes called the "localization method". An important part of this theory is to give conditions under which an abstract limit functional has a representation as an integral functional. The atomistic energies are not integral functionals and can not be represented as such. Therefore, the existing theory does not apply directly even if one assumes the correct growth conditions. Nevertheless, in [AC04] Alicandro and Cicalese adapt the method to the context of discrete-to-continuum limits and prove a general integral representation result for continuum limits of atomistic pair interaction potentials. Departing from that, we will derive a continuum theory for very general interaction potentials which, in particular, can also incorporate bond-angle dependent potentials and other, more complex, multi-body interactions. Such an extension is desirable in applications, as many atomistic models such as, e.g., the Stillinger-Weber potential, cannot be written as a pure pair potential. In fact, the class of potentials our theory applies to is rich enough to model any quasiconvex continuum energy density with the correct growth conditions which in particular includes every possible linearized elasticity tensor, even if the Cauchy-relations are violated. First steps in this direction have been provided through the analysis of a special class of nearest neighbor three point interactions on a two-dimensional square lattice by Meunier, Pantz and Raoult, see [MPR12], and by Alicandro, Cicalese and Gloria–in the more general context of stochastic lattices–for pair interactions with an additional term penalizing volume changes, see [ACG11]. We wish to emphasize that, in contrast, our approach allows for full finite range many body interactions.

The limiting energy density will be described in terms of a sequence of cell problems. This is related to the homogenization results of Braides [Bra85] and Müller [Mü87] for non-convex integral functionals. This representation will then allow us to discuss the applicability of the results in [CDKM06] about the Cauchy-Born rule mentioned above. Indeed we will show that under certain additional assumptions the limiting energy density is indeed given by the Cauchy-Born energy density for small strains.

For such results to be realistic, it is clear that one has to look at models of materials that behave elastically globally. In particular, the material must behave in a elastic way for arbitrarily large deformations. Mathematically, this is ensured by a strong growth condition on the atomic interaction potentials. While these result are mathematically quite pleasing and fit well into already established theory, they are quite unsatisfactory from a more applied point of view. Typical interaction potentials in atomistic modeling, e.g., the Lennard-Jones potential, do not satisfy such growth conditions. On the contrary, they typically become almost constant for large deformations which reflects the fact that atoms which are far away from each other do not significantly interact anymore. If one then considers even simple boundary conditions, e.g. an affine extension of the material, the global minimizer of the energy no longer behaves purely elastically and–among other things–might develop cracks (cf. [FS14]).

Nevertheless, a lot of actual materials still behave elastically at least for a certain range of finite deformations. Physically, there is no need for a stable state to be global energy minimizer. Indeed, based on the observation that elastic deformations in general are merely local energy minimizers, E and Ming have pioneered a different approach. In [EM07b] and [EM07a] they consider the formal asymptotic expansion of the atomistic equations in $\varepsilon$. Writing $F_\varepsilon(y) = 0$ for the atomistic equations and $F(y) = 0$ for the Cauchy-Born continuum equations (see Chapter 2) where $y$ is the deformation mapping, one formally finds $F_\varepsilon(y) = F(y) + O(\varepsilon^2)$ in the case of a Bravais lattice. If this is turned into a rigorous residual estimate and one also establishes control on the stability and certain differentiability and continuity properties, then one can hope to apply a fixed point theorem (or an inverse/implicit function theorem depending on the precise situation), to prove the existence of atomistic solutions close to a solution of the Cauchy-Born continuum problem. It turns out that a certain crucial continuity constant that needs to be controlled in this approach actually blows up as $\varepsilon \to 0$. To overcome such problems and also to provide explicit control over the dependence on $\varepsilon$ it is crucial to work with a quantitative version of the fixed point theorem.

E and Ming use this idea to show that under certain stability assumptions, smooth solutions of the equations of continuum elasticity on the flat torus are asymptotically approximated by corresponding atomistic configurations. More recently these results have been clarified and generalized to remarkably mild regularity assumptions for problems on the whole space with a far field condition by Ortner and Theil, cf. [OT13].

In view of these results, the natural question arises if an analogous analysis is possible for a material occupying a general finite domain in space on the boundary of which there might also be prescribed boundary values. To cite Ericksen [Eri08, p. 207], "Cannot someone do something like this for a more realistic case, say zero surface tractions on part of the boundary and given displacements on the remainder?" This is what we will indeed do in the main part of this thesis. While we formulate our results only in the case of fixed displacements on the entire boundary, traction and mixed boundary conditions are to a large part implicitly included. While one would typically need slightly stronger stability assumptions to ensure existence and smoothness of solutions to the continuum problem under mixed or Neumann boundary conditions, once these properties are established one can just view the solution as a solution under Dirichlet boundary values. And if one has a solution to the atomistic equations under given Dirichlet boundary conditions on a set of boundary atoms, one can just as easily declare these boundary atoms as non-physical ghost atoms that generate certain forces in their interaction range. Thus we also have a solution under a (specific) traction boundary condition. Either way, while we will treat more or less arbitrary continuous boundary data, our results are restricted in that we can only consider a certain range of atomistic boundary data corresponding to the continuum boundary data in the limit. But this restriction actually is to be expected. Indeed, as we will argue below, in the case of general atomistic boundary data the Cauchy-Born rule is typically expected to fail due to relaxation effects at the boundary.

This treatment of arbitrary domains and general displacement boundary conditions is of interest not only from a theoretical perspective but also with a view to specific domains and boundary constraints that are of interest in applications, where one can use our results as a starting point for further analysis. Besides discussing the elastic behavior, one can also treat questions of stability and even the onset of instabilities, since the stability assumptions we make are designed to be sharp.

In order to relate our set-up to the aforementioned previous contributions on the full space and the flat torus, we remark that the presence of displacement boundary conditions leads to some subtleties within the statement of our main theorems, Theorem 4.5 and Theorem 5.8, and to a number of challenges and technical difficulties within its proof: 1. The boundary values are naturally imposed on a boundary layer of the atomistic system. In contrast to the situation, e.g., in [CDKM06], the adequate choice of the atomistic displacements at the boundary for a general non-affine continuum boundary datum is not determined canonically a priori. Instead, one needs to construct the correct atomistic displacements from the continuum Cauchy-Born solution. Doing so we see that not only the correct asymptotic continuous boundary values but also the correct asymptotic normal derivative given by the normal derivative of the continuous Cauchy-Born solution are attained. (If these conditions fail, we again typically expect surface relaxation effects and a failure of the Cauchy-Born rule close to the boundary.) In the dynamic case one additionally has to control the time dependency of the boundary values. 2. In order to allow for as many atomistic boundary conditions (and body forces) as possible, we consider general scalings $\varepsilon^\gamma$ in our theorems and only restrict $\gamma$ as much as necessary (basically $\frac{d}{2} < \gamma \leq 2$). While smaller $\gamma$ will lead to a larger variety of atomistic boundary values, $\gamma = 2$ will lead to the optimal convergence rates. One should also note that our result in the static case no longer requires $\varepsilon$ to be small in contrast to previous results. 3. Certain technical methods, which are available on the flat torus or on the whole space, do not translate to our setting. E.g., quasi-interpolations as in [OS12] do not preserve boundary conditions. This leads us to prove the important residual estimates, which lie at the core of our main proof, in a different and more robust way. With the help of a subtle atomic scale regularization, this can be achieved by requiring only slightly higher regularity assumptions for the continuous equations as compared to [OT13] in the static case. In the dynamic problem we actually do not need any additional regularity as the regularity assumptions in the relevant case of the dynamic result in [OT13] are higher as well. 4. The time-dependent boundary conditions lead to additional error terms in the crucial energy estimates for our dynamic result. Inspired by previously developed estimates based on one or two partial integrations in time (cf. [OT13]) that were designed to deal with the nonlinearity, we introduce a more general scheme based on arbitrary many partial integrations in time that allows us not only to reduce the assumptions on the convergence rates as mentioned above, but that is also applicable to the additional error terms from the boundary.

Let us discuss in a little bit more detail the still open problem of general atomistic boundary values that do not satisfy our constraints. While in the bulk it is plausible that the Cauchy-Born rule is still approximately true, the situation is–as mentioned–very different close to the surface. In the outermost layers one expects surface relaxation effects. E.g., in case of a free boundary one expects the gradients to have an error of $\mathcal{O}(1)$ while oscillating on the scale $\mathcal{O}(\varepsilon)$. Even though this does not effect the highest order of the energy, it does mean that the Cauchy-Born approximation leaves a residual in the equations that does not vanish as $\varepsilon \to 0$, e.g., in any $L^p$-norm. This makes it much more difficult to find exact solutions to the equations with asymptotically equal bulk behavior. A precise and rigorous mathematical treatment of surface relaxation effects in more than one dimension is currently still out of reach. The best known result so far appears to be [The11], which gives the correct asymptotics of the surface energy in the limit of vanishing mismatches in the potentials. But even if one were to establish a full characterization of the surface energy, this would still be just a first step towards describing exact solutions of the equations.

A major assumption in all of the local results discussed up to here is that the deformations must be sufficiently close to a stable affine lattice. The precise required stability assumption, the so called atomistic stability, was first discussed in detail in a recent work by Hudson and Ortner, cf. [HO12]. Building on these results, we will discuss the stability condition in the case of a finite domain with boundary conditions, prove a new and more intuitive characterization, give new and easy to check criteria, and discuss two examples analytically. Probably the most important insight in this discussion is that the stability of the continuum problem is insufficient for the stability of the atomistic problem, cf. Section 3.3 for a counterexample.

In the already mentioned articles [EM07b], [EM07a], and [OT13] the authors not only consider the equilibrium situation but also discuss the dynamic equations on the torus and the full space, respectively. We want to do the same on a finite domain where we additionally have to take into account fully time-dependent boundary conditions. While equilibrium situations play an important role in mechanics, in many situations the material behavior decisively depends on inertial effects and, as a consequence, static or quasistatic descriptions are insufficient. Mathematically, the dynamic equations are considerably more challenging. Already in the continuum description, we have to discuss a quasi-linear, second-order, hyperbolic system under time-dependent boundary conditions. While the problems have natural energies that are conserved, they are of limited use in the analysis since level sets are typically unbounded in relevant norms and can contain regions where there is a loss of stability.

Short time results on the equations of continuum nonlinear elastodynamics already exist for quite some time. Let us just mention [HKM77] by Hughes, Kato, and Marsden which pursues an abstract approach, as well as [DH85] by Dafermos and Hrusa which provides similar results via more direct norm estimates. Not only do we want to adapt these existing results to our equations, we also want to extend and characterize the maximal existence time.

For the atomistic equations of elastodynamics it is easy to establish existence of solutions close to the continuum solution for short times. The actually interesting and difficult problem is to show that existence and convergence hold for a macroscopic time interval that does not vanish as $\varepsilon \to 0$. Indeed, we no longer require any kind of fixed point or implicit function argument. Instead, the key point is to establish good estimates on certain norms of the difference between the solutions.

Despite this very different approach, the above mentioned residual estimates, that make the formal asymptotic expansion rigorous, still play a major role.

In comparison to previous works, besides considering a finite domain with boundary conditions, there is an additional open question of considerable importance. Is it possible to establish the existence of atomistic solutions that satisfy the Cauchy-Born rule and their link to the continuum solutions not just for small times and deformations close to a stable affine configuration but for long times and large deformations? In [EM07a], E and Ming only proof a short time result. An extension is not obvious, since their methods were restricted to small displacements. In [OT13], Ortner and Theil are indeed aware of this restriction that also applies to their results. They proposed that one could indeed extend the results to long times if one were to establish an atomistic version of the Gårding inequality.

The need for such an inequality can be understood as follows. For small displacements it is sufficient to stay close enough to a given stable affine deformation to ensure that the second variation of the potential energy is positive uniformly in $H_0^1$. For large deformations this is, in general, false. Even if at each point the gradient corresponds locally to a stable affine deformation, the second variation can still be negative globally. A Gårding inequality helps to work around this situation. It states that one can still get uniform positivity in $H_0^1$ from the local stability if one is willing to add a large constant times a lower order term (more precisely, the square of the $L^2$-norm, cf. Theorem C.1). While a Gårding inequality alone is not sufficient to treat large deformations for the static case, this inequality is key to the dynamic equations. In the continuum case the Gårding inequality is indeed part of a well-established theory. In the discrete case such a Gårding inequality is more subtle. Already the question of what constitutes a 'locally stable deformation' requires–as mentioned–a deeper analysis in the atomistic case, which is the main topic of Chapter 3. In particular, the local stability assumption from the continuum case turns out to be insufficient. Additionally, in the continuous case the continuity of the coefficients and the deformation gradient is a crucial assumption for the Gårding inequality. This assumption has to be replaced by a more quantified version that is adapted to the discrete nature of the atomistic problem. In this spirit we will indeed establish a new atomistic Gårding inequality, Theorem 5.6.

Being able to control large deformations, we can then prove the existence and convergence result for long times. More precisely, Theorem 5.8 states that as long as the solution of the continuum Cauchy-Born problem exists and is atomistically stable everywhere there are solutions to the atomistic problems close by and they converge to the continuum solution as $\varepsilon \to 0$. Again, we give precise conditions on the choice of atomistic boundary values, initial conditions, and body forces.

Having already discussed two very different approaches to the Cauchy-Born rule and the convergence result for solutions, we want to conclude the thesis with a third and new approach for the static case in Chapter 6. Again we will not assume any growth conditions and work in a local setting. We will also require much lower regularity assumptions than in our static results in Chapter 4. This comes at the price that we can not say anything about the existence of atomistic solutions in the range we discuss. Still, if we start from atomistic solutions that are sufficiently close to a stable affine map with only a condition on the convergence of the boundary data, we are able to show the convergence of the entire sequence strongly in $H^1$. We can also show that the limit is a solution of the continuum problem and that the energies converge. The strong convergence of the gradients (in integral norms) indeed is a new type of justification of the Cauchy-Born rule for the bulk behavior. The deformations in any sequence of atomistic solutions along a vanishing sequence of interatomic distances, subject to conditions at the boundary, do indeed follow the gradient of a continuum solution in the bulk. And indeed the Cauchy-Born energy density gives the correct limiting energy and, more importantly, the correct limiting equations. The proof is mainly inspired by variational methods from the theory of $\Gamma$-convergence and ideas about uniformly monotone operators. Of course, we actually discuss local and not just global minimizers and the first variation is typically not a monotone operator. Therefore, we have to implement these ideas in a completely local approach.

# Outline

Let us give a short outline of the thesis.

As discussed we want to start in the fist part in Chapter 1 by considering a global approach to the Cauchy-Born rule and the discrete-to-continuum limit. Besides our main $\Gamma$-convergence and compactness results (Theorem 1.1, 1.3, 1.5, 1.7, 1.8), we discuss the convergence of (almost) minimizers (Corollary 1.6), the applicability of the Cauchy-Born rule (Theorem 1.9), as well as the validity of the Cauchy relations for the linearization. As part of our proof we also improve existing integral representation results from the quasiconvex case (Theorem 1.16) to the cross-quasiconvex case (Theorem 1.30).

The main part of this thesis is the second part. Here, we discuss and prove the local results mentioned in the introduction that do not require any growth conditions.

We start in Chapter 2 with a thorough discussion of the models we consider and the basic assumptions we make.

In Chapter 3 we discuss the crucial stability concepts in detail. We define the stability constants and give some basic properties in Section 3.1, establish their main characterizations in Fourier variables (Theorem 3.7, Corollary 3.9) in Section 3.2, and give examples that highlight important properties in Section 3.3.

Chapter 4 is devoted to the static case. We begin with the continuum problem in Section 4.1. Then we state our main existence and convergence result for the static case in Section 4.2, namely Theorem 4.5. The rest of the chapter is devoted to the proof. It is based on a quantitative implicit function theorem (Theorem 4.7, Corollary 4.8). In the proof we will, in particular, make the formal expansion of the atomistic equations in $\varepsilon$ precise as we prove rigorous residual estimates in Section 4.3.

Then, in Chapter 5, we treat the dynamic case. Again, we start with the continuum problem in Section 5.1. The main result here, Theorem 5.4, discusses the maximal existence interval of the continuum solution. To get there we first discuss the compatibility conditions in detail and establish short time existence. In Section 5.2, we state and prove the atomistic Gårding inequality, Theorem 5.6, as well as our main existence and convergence result in the dynamic case, Theorem 5.8.

As mentioned in the introduction, we want to complement the results from Chapter 4 with a pure convergence result under weaker assumptions. This is done in Chapter 6.

At last, in the appendices we collect and prove a few technical lemmata concerning elliptic and hyperbolic regularity as well as the multiplication of many Sobolev functions which we will mostly need in Chapter 5. Some of these results might already be (implicitly) known to experts in the field but do not seem to be available in the literature. Some other results will be directly quoted from the existing literature.

Given this outline, let us also clearly state the relation of the different chapters of this thesis to previously published articles of the author as well as manuscripts that will be published in the near future.

The contents of the first chapter have been published in [BS13]. This peer-reviewed article was also accepted as the master's thesis of the author. Large parts of it also overlap with the bachelor's thesis of the author. Nevertheless, it is still important to be included in this thesis to show a completely different approach to the Cauchy-Born rule and the convergence result.

The more important part of this thesis is the second part. Here, the largest part of Chapters 2, 3, 4, 5 (but not Chapter 6), as well as most of the appendices is currently submitted for peer-reviewed publication and is already published in the preprints [BS16], [Bra16].

# Acknowledgments

I would like to express my sincere gratitude to my advisor Bernd Schmidt, who has supported me with his patience and knowledge whilst allowing me the room to work in my own way. I also very much appreciate that Lisa Beck, Dirk Blömker, Christoph Ortner, and Florian Theil were all willing to take some time to discuss different parts of this thesis with me.

Furthermore, I would like to thank all the people in the research group "Nichtlineare Analysis" at the Universität Augsburg for making sure that it never got boring. In particular, thanks go Martin Jesenko for putting up with me for all this time. Additionally, I would like to show my gratitude to the M.O.P.S. for making the daily grind more enjoyable.

Last but not least, I would like to thank my family for supporting me throughout.

# Part I

# A First Approach: The Global Minimization Problem under Additional Growth Conditions

# Chapter 1

# A Γ-Convergence Result for Potentials with $p$-Growth

## 1.1  Introduction and Results

The main aim of this chapter is to provide a rigorous derivation of continuum elasticity functionals from atomistic models in terms of Γ-convergence for atomistic models that allow for general finite range interactions under certain growth assumptions.

As stated in the outline, the content of this chapter has been published in [BS13] which was accepted as the master's thesis of the author and also has a strong overlap with the bachelor's thesis of the author.

In order to prove the main representation result in this chapter we resort to abstract compactness properties of Γ-limits and integral representation results for functionals on Sobolev spaces and thus follow the scheme, which is dictated by verifying the hypotheses of that abstract approach by the localization method. Some of the arguments in this proof are suitable adjustments of the corresponding results in [AC04]. There are, however, some major differences as compared to the pair interaction case treated by Alicandro and Cicalese. While these authors use slicing arguments in order to obtain energy estimates on the usual $d \times d$ deformation gradients in the direction of interacting pairs, we will have to estimate the much higher dimensional $d \times N$, $N \geq 2^d$, discrete deformation gradients. In fact, as in general our discrete energies cannot be recovered by slicing techniques, we will instead work with interpolations of the discrete deformations which encode the full discrete gradient on lattice cells.

Let us discuss the precise setting for this part of the thesis. We first consider systems with short range interactions on a simple lattice: If $\mathcal{L} = A\mathbb{Z}^d$ is some Bravais lattice, $\Omega \subset \mathbb{R}^d$ a bounded open set with Lipschitz boundary that will be viewed as the 'macroscopic' domain occupied by the elastic body, whose atoms are at positions $\varepsilon\mathcal{L} \cap \Omega$, we assume that the energy of a deformation $y : \varepsilon\mathcal{L} \cap \Omega \to \mathbb{R}^d$ can be written in the form

$$F_\varepsilon(y, \Omega) = \varepsilon^d \sum_{x \in (\mathcal{L}'_\varepsilon(\Omega))^\circ} W_{\text{cell}}(\bar{\nabla}y(x)),$$

where $x$ runs over all midpoints of elementary lattice cells of $\varepsilon\mathcal{L}$ inside $\Omega$. Here $\bar{\nabla}y(x)$ is the discrete gradient of $y$ on the corresponding cell $Q$ which encodes all relative displacements of atoms lying on the corners of $Q$. (See Section 1.2 for precise definitions.) $\varepsilon$ is the small parameter in the system measuring the typical interatomic distance and tending to zero eventually in the continuum limit. The rescaling by $\varepsilon^d$ is introduced in order to pass from units of finite energy per atom to units of finite energy per unit volume.

In fact, our analysis is not restricted to interactions within unit lattice cells, but also applies to general finite range interactions. In such models, the energy is still given as the sum over unit lattice cells $(\mathcal{L}'_\varepsilon(\Omega))^\circ$, but the cell energy now depends on the discrete deformation gradient $\bar{\nabla}y(x) \in \mathbb{R}^{d \times N}$ on a larger 'super-cell':

$$F_\varepsilon(y, \Omega) = \varepsilon^d \sum_{x \in (\mathcal{L}'_\varepsilon(\Omega))^\circ} W_{\text{super-cell}}(\bar{\nabla}y(x)),$$

where the definition of the lattice interior $(\mathcal{L}'_\varepsilon(\Omega))^\circ$ is suitably adjusted so that only lattice points within $\Omega$ may interact.

Another complication arises when extending our results to general finite-range interactions on multi-lattices. For such systems, the discrete gradients are augmented with additional internal variables describing the relative shifts of the underlying single lattice. With the help of a mixed Sobolev/Lebesgue space representation theorem we are then led to a boundary value/mean value cell formula for the limiting energy density. This cell formula is in fact related to the cell formula derived in [Sch08] for thin membranes where internal variables measure relative shifts of the thin film's layers. On multi-lattices $\varepsilon(\{0, s_1, \ldots, s_m\} + \mathcal{L})$ with $m$

shift vectors $s_1, \ldots, s_m \in \mathbb{R}^d$ the general discrete energy functional then reads

$$F_\varepsilon(y, s, \Omega) = \varepsilon^d \sum_{x \in (\mathcal{L}'_\varepsilon(\Omega))^\circ} W_{\text{super-cell}}(\bar{\nabla} y(x), s(x)),$$

where $y : \varepsilon \mathcal{L} \cap \Omega \to \mathbb{R}^d$, $s : \varepsilon \mathcal{L} \cap \Omega \to \mathbb{R}^{d \times m}$. For notational convenience we will restrict to simple unit cell interactions on a Bravais lattice for the largest part of the paper and only comment on the necessary modifications in the more general case at the end of Section 1.5.

Our main results are summarized in the following theorems. The Assumptions 1, 2 and 3 on the cell energy are specified in Section 1.2. (Assumptions 1 and 2 are nothing but standard $p$-growth assumptions on $W_{\text{cell}}$.)

**Theorem 1.1** (Γ-convergence). *Suppose Assumptions 1 and 2 are true. Then $F_\varepsilon(\cdot, \Omega)$ $\Gamma(L^p(\Omega; \mathbb{R}^d))$- and $\Gamma(L^p_{loc}(\Omega; \mathbb{R}^d)/\mathbb{R})$-converges to the functional $F$, defined by*

$$F(y) = \begin{cases} \displaystyle\int_\Omega W_{\text{cont}}(\nabla y(x)) \, dx, & \text{if } y \in W^{1,p}(\Omega; \mathbb{R}^d), \\ \infty & \text{otherwise,} \end{cases}$$

*where the continuum density $W_{\text{cont}} \colon \mathbb{R}^{d \times d} \to [0, \infty)$ is given in terms of $W_{\text{cell}}$ by*

$$W_{\text{cont}}(M) = \frac{1}{|\det A|} \lim_{N \to \infty} \frac{1}{N^d} \inf \left\{ \sum_{x \in (\mathcal{L}'_1(A(0,N)^d))^\circ} W_{\text{cell}}(\bar{\nabla} y(x)) : \right.$$
$$\left. y \in \mathcal{B}_1(A(0, N)^d, y_M) \right\}.$$

Here $\mathcal{B}_1(A(0, N)^d, y_M)$ is the space of lattice deformations of $\mathcal{L}_1 \cap A(0, N)^d$ with linear boundary conditions $M$ on $\partial \mathcal{L}_1(A(0, 1)^d)$, cf. Section 1.4.

*Remark* 1.2. As in non-convex homogenization (see [Mü87]), in the representation result for $W_{\text{cont}}$ it is necessary to minimize $W_{\text{cell}}$ over larger and larger cubes and the limit is in general not obtained for finite $N$. A simple 2d example for this effect is given by a square lattice where nearest neighbor atoms interact via a harmonic spring potential: $F_\varepsilon(y) = \frac{1}{2} \sum_{|x-x'|=\varepsilon} (|y(x) - y(x')| - \varepsilon)^2$ (which can be written in the above form). This is a simplified version of the example discussed in [Sch08, Sect. 4.4]. The arguments sketched there, which amount to considering deformations

$$y(x_1, x_2) = M(x_1, x_2) + \sigma_1(x_1) + \sigma_2(x_2)$$

for compressive boundary conditions, with suitable 2-periodic functions $\sigma_1$ and $\sigma_2$ like

$$\sigma_i(z) = \frac{1}{2}(-1)^z \sqrt{\frac{1}{m_{1i}^2 + m_{2i}^2} - 1} \begin{pmatrix} -m_{2i} \\ m_{1i} \end{pmatrix},$$

$i = 1, 2$, and suitably modified on the boundary, show that

$$\frac{1}{N^2} \inf \left\{ \sum_{x \in (\mathcal{L}_1'((0,N)^d))^\circ} W_{\text{cell}}(\bar{\nabla} y(x)) : y \in \mathcal{B}_1((0,N)^d, y_M) \right\}$$

converges to $(\max\{0, |m_{.1}| - 1\})^2 + (\max\{0, |m_{.2}| - 1\})^2$ with error bound $O(N^{-1})$, where $m_{.j}$ denotes the $j$th column of $M$. Evaluating, however, the energy of the bonds on and close to the boundary it is easily seen that, for compressive boundary conditions $|m_{.1}| < 1$ or $|m_{.2}| < 1$, this limiting energy is always over-estimated by a constant times $N^{-1}$.

**Theorem 1.3** (Compactness). *Under the assumptions of Theorem 1.1, if $y_\varepsilon$ is a sequence with equibounded energies $F_\varepsilon(y_\varepsilon, \Omega)$ and $\Omega$ is connected, then there exist a sequence $\varepsilon_k \to 0$ and $y \in W^{1,p}(\Omega; \mathbb{R}^d)$ such that $y_{\varepsilon_k} \to y$ in $L^p_{loc}(\Omega; \mathbb{R}^d)/\mathbb{R}$.*

Of course, if $\Omega$ is not connected, one has compactness in $L^p_{loc}$ up to translation on every connected component.

Analogous results hold true if one adds boundary conditions $g \in W^{1,\infty}(\mathbb{R}^d; \mathbb{R}^d)$. Let $F^g$ and $F^g_\varepsilon$ denote the functionals obtained from $F$ and $F_\varepsilon$, respectively, with values set to infinity if the boundary conditions are not met, cf. Section 1.4.

**Theorem 1.4** (Γ-convergence). *If Assumptions 1 and 2 are true, then $F^g_\varepsilon(\cdot, \Omega)$ Γ$(L^p(\Omega; \mathbb{R}^d))$-converges to the functional $F^g$.*

**Theorem 1.5** (Compactness). *Under the assumptions of Theorem 1.4, if $y_\varepsilon$ is a sequence with equibounded energies $F^g_\varepsilon(y_\varepsilon, \Omega)$, then there exist a sequence $\varepsilon_k \to 0$ and $y \in W^{1,p}(\Omega; \mathbb{R}^d)$ with $y = g$ on $\partial\Omega$ such that $y_{\varepsilon_k} \to y$ in $L^p(\Omega; \mathbb{R}^d)$.*

A standard argument then yields that almost minimizers of $F_\varepsilon^g(\cdot, \Omega)$ converge to minimizers of $F^g(\cdot, \Omega)$ and almost minimizers of $F_\varepsilon(\cdot, \Omega)$ up to translation converge to minimizers of $F(\cdot, \Omega)$, more precisely:

**Corollary 1.6** (Convergence of almost minimizers). *Suppose Assumptions 1 and 2 are true. Then every sequence of almost minimizers of $F_\varepsilon(\cdot, \Omega)$ for connected $\Omega$ is compact in $L_{loc}^p(\Omega; \mathbb{R}^d)/\mathbb{R}$ and every limit is a minimizer of $F$, while every sequence of almost minimizers of $F_\varepsilon^g(\cdot, \Omega)$ is compact in $L^p(\Omega; \mathbb{R}^d)$ and every limit is a minimizer of $F^g$.*

It is not hard to include body forces in the energy expression as these will only be continuous perturbations of the energy functional which converge uniformly on bounded sets and thus preserve $\Gamma$-convergence by the general theory.

We also remark that the point why the theory can be adapted to the case of general finite range interactions, is that in this case $W_{\text{cell}}$, while naturally still bounded from above by the discrete gradient through Assumption 2, from below has to be bounded only in terms of the discrete gradient on the unit cell. See Section 1.5 for details. For general finite range interactions on multi-lattices we state here only the analogue of Theroem 1.1, as in the cell formula there are now additional internal variables that need to be taken into account. Theorems 1.3, 1.4 and 1.5 and Corollary 1.6 extend in a straightforward manner.

**Theorem 1.7** ($\Gamma$-convergence). *Suppose $W_{\text{super-cell}}$ satisfies the growth assumptions (as stated in Section 1.5). Then $F_\varepsilon(\cdot, \cdot, \Omega)$ $\Gamma(L^p(\Omega; \mathbb{R}^d) \times w\text{-}L^q(\Omega; \mathbb{R}^{d \times m}))$-converges to the functional $F$, defined by*

$$F(y, s) = \begin{cases} \int_\Omega W_{\text{cont}}(\nabla y(x), s(x)) \, dx, & \text{if } y \in W^{1,p}(\Omega; \mathbb{R}^d), \\ \infty & \text{otherwise,} \end{cases}$$

*where the continuum density $W_{\text{cont}} \colon \mathbb{R}^{d \times d} \times \mathbb{R}^{d \times m} \to [0, \infty)$ is given in terms of $W_{\text{super-cell}}$ by*

$$W_{\text{cont}}(M, s_0) = \frac{1}{|\det A|} \lim_{N \to \infty} \frac{1}{N^d} \inf \Bigg\{ \sum_{x \in (\mathcal{L}_1'(A(0,N)^d))^\circ}$$

$$W_{\text{super-cell}}(\bar\nabla y(x), s(x)) \colon (y, s) \in \mathcal{B}_1(A(0, N)^d, y_M, s_0) \Bigg\}.$$

Here $\mathcal{B}_1(A(0, N)^d, y_M, s_0)$ is the space of lattice deformations of $\mathcal{L}_1 \cap A(0, N)^d$ with linear boundary conditions $M$ on $\partial \mathcal{L}_1(A(0, 1)^d)$ for $y$ and average $s_0$ for $s$, cf. Section 1.5.

As macroscopic deformations are solely given in terms of a deformation mapping $y \in W^{1,p}(\Omega; \mathbb{R}^d)$, we are also interested in the effective macroscopic energy density obtained by minimizing out the internal variables $s$:

**Theorem 1.8.** *For every* $y \in W^{1,p}(\Omega; \mathbb{R}^d)$ *we have*

$$\min_{s \in L^q(\Omega; \mathbb{R}^{d \times m})} \int_\Omega W_{\mathrm{cont}}(\nabla y(x), s(x))\, dx = \int_\Omega \min_{s \in \mathbb{R}^{d \times m}} W_{\mathrm{cont}}(\nabla y(x), s)\, dx.$$

*Moreover,* $F_\varepsilon^{s-\min}(\cdot, \Omega) := \inf_{s \in L^q} F_\varepsilon(\cdot, s, \Omega)$ $\Gamma(L^p(\Omega; \mathbb{R}^d))$*-converges to the functional* $F^{s-\min}$*, defined by*

$$F^{s-\min}(y) = \begin{cases} \displaystyle\int_\Omega \min_{s \in \mathbb{R}^{d \times m}} W_{\mathrm{cont}}(\nabla y(x), s)\, dx, & \text{if } y \in W^{1,p}(\Omega; \mathbb{R}^d), \\[2mm] \infty & \text{otherwise,} \end{cases}$$

Returning to our basic setting on a Bravais lattice, under an additional assumption, we can calculate the limiting density for small strains explicitly by the Cauchy-Born rule:

**Theorem 1.9.** *In addition to Assumptions 1 and 2 suppose that* $W_{\mathrm{cell}}$ *satisfies Assumption 3. Then there is a neighborhood* $\mathcal{U}$ *of* $SO(d)$ *such that* $W_{\mathrm{cont}}$ *is given by*

$$W_{\mathrm{cont}}(M) = W_{\mathrm{CB}}(M) := \frac{1}{|\det A|} W_{\mathrm{cell}}(MZ).$$

*for all* $M \in \mathcal{U}$.

Here $Z \in \mathbb{R}^{d \times 2^d}$ is a 'discrete identity matrix', see Section 1.2 for details.

As $W_{\mathrm{cont}}$ arises as the energy density of a Γ-limit it has to be quasiconvex (cf. Section 1.2 for the definition of these concepts). The next proposition, which in fact is a direct consequence of the convergence of the P1 finite element method in nonlinear elasticity, shows that our class of atomistic interactions is rich enough to model any quasiconvex energy density in the continuum limit.

**Proposition 1.10.** *Suppose $V : \mathbb{R}^{d \times d} \to \mathbb{R}$ is quasiconvex with standard p-growth*

$$c|M|^p - c' \leq V(M) \leq c''(|M|^p + 1)$$

*for some constants $c, c', c'' > 0$ and all $M \in \mathbb{R}^{d \times d}$. Then there exists a cell energy $W_{\mathrm{cell}}$ satisfying Assumptions 1 and 2 such that $W_{\mathrm{cont}} = V$.*

We remark that, by way of contrast, a restriction to pair interaction models will only lead to a restricted class of limiting continuum energies, as can be quantified in terms of the so-called Cauchy relations: If the Cauchy-Born rule applies (e.g., due to Assumption 3), an atomistic interaction energy

$$E(y) = \varepsilon^d \sum_{\substack{x, x' \in \mathcal{L}_\varepsilon \cap \Omega \\ x \neq x'}} V_{\frac{|x - x'|}{\varepsilon}} \left( \frac{|y(x) - y(x')|}{\varepsilon} \right)$$

yields the continuum density

$$W_{\mathrm{CB}}(M) = \frac{1}{|\det A|} \sum_{\substack{x \in \mathcal{L} \\ x \neq 0}} V_{|x|}(|Mx|).$$

Assuming $V_{|x|}$ is smooth and, for large $|x|$, sufficiently rapidly decreasing a direct calculation yields

$$D^2 W_{\mathrm{CB}}(\mathrm{Id})(M, M) = \sum_{i,j,k,l=1}^{d} c_{ijkl} m_{ij} m_{kl},$$

where the elastic constants $c_{ijkl}$ are given by

$$c_{ijkl} = \frac{1}{|\det A|} \sum_{\substack{x \in \mathcal{L} \\ x \neq 0}} V_{|x|}''(|x|) \frac{x_i x_j x_k x_l}{|x|^2} + V_{|x|}'(|x|) \left( \frac{x_j x_l \delta_{ki}}{|x|} - \frac{x_i x_j x_k x_l}{|x|^3} \right).$$

While the symmetry relations $c_{ijkl} = c_{klij}$ and $c_{ijkl} = c_{jikl}$ naturally follow from the symmetry of the Hessian $D^2 W_{\mathrm{CB}}(\mathrm{Id})$ and frame indifference of $W_{\mathrm{CB}}$, the particular form of $W_{\mathrm{CB}}$ in addition gives $c_{ijkl} = c_{ilkj} = c_{kjil}$ for every $i, j, k, l$.

In the 3-dimensional setting of elasticity theory these additional relations lower the dimension of admissible elasticity tensors from 21 to

15 (symmetric in all indices) and so can be written as 6 equations, the Cauchy-relations, namely

$$c_{1122} = c_{1212}, \quad c_{2233} = c_{2323}, \quad c_{3311} = c_{3131},$$
$$c_{1123} = c_{1213}, \quad c_{2231} = c_{2321}, \quad c_{3312} = c_{3132}.$$

The question whether in elasticity theory the Cauchy-relations hold true (rari-constant theory) or fail (multi-constant theory) had been under discussion for quite some time in physics and was finally decided by experimental data in favor of the multi-constant theory (some experimental data and further physical considerations can be found, e.g., in [Hau67]). This means, in particular, that the interaction in a lattice is a complex multibody interaction which cannot be reduced to pairpotentials. Our model in this paper using general cell energies is not limited by the Cauchy-relations:

**Proposition 1.11.** *Suppose* $Q : \mathbb{R}^{d \times d} \to \mathbb{R}$ *is a positive semidefinite quadratic form which is positive definite on the symmetric* $d \times d$ *matrices and vanishes on antisymmetric matrices. Then there exists* $W_{\mathrm{cell}}$ *satisfying Assumptions 1, 2 and 3 such that*

$$\frac{1}{2}D^2 W_{\mathrm{CB}}(\mathrm{Id})(M, M) = Q(M).$$

The paper is organized as follows. In Section 1.2 we first introduce the discrete model and review some basic facts on Γ-convergence and integral representations of functionals on Sobolev spaces. In Section 1.3 we then proceed to state precisely and prove a general Γ-compactness and representation theorem. This in particular requires a number of technical preliminaries in order to investigate discrete deformations. A version of this representation result for boundary value problems is then provided in Section 1.4. Finally, the limiting stored energy function is identified in Section 1.5 through minimizing a sequence of cell problems, leading to the main discrete-to-continuum convergence result and the proofs of the results stated above.

## 1.2   The Model and Preliminaries

In this section we introduce the atomistic model and recall some general facts on Γ-convergence and integral representation results required by the localization method.

## The Atomistic Model

Let $\mathcal{L} \subset \mathbb{R}^d$ be a Bravais lattice, i.e., there are linearly independent vectors $v_1, \dots, v_d$ such that

$$\mathcal{L} = \{n_1 v_1 + \cdots + n_d v_d \mid n_1, \dots, n_d \in \mathbb{Z}\} = A\mathbb{Z}^d,$$

if we set $A = (v_1, \dots, v_d)$. The scaled lattices $\mathcal{L}_\varepsilon = \varepsilon\mathcal{L}$ partition $\mathbb{R}^d$ into the $\varepsilon$-cells $z + A[0, \varepsilon)^d$ $(z \in \mathcal{L}_\varepsilon)$. Let $Q_\varepsilon(x)$ denote the $\varepsilon$-cell containing $x$. The centers of the cells are $\mathcal{L}'_\varepsilon = \mathcal{L}_\varepsilon + A(\frac{1}{2}, \dots, \frac{1}{2})$ and we denote by $\bar{x}$ the center of the cell containing $x$. These centers give a convenient labeling of the cells. Furthermore let $z_1, \dots, z_{2^d}$ be the points in $A\left\{-\frac{1}{2}, \frac{1}{2}\right\}^d$ and $Z := (z_1, \dots, z_{2^d}) \in \mathbb{R}^{d \times 2^d}$.

For a set $U \subset \mathbb{R}^d$ we define the following lattice subsets in the spirit of its closed hull, interior or boundary with respect to $\varepsilon\mathcal{L}'$ or its corners $\varepsilon\mathcal{L}$ by

$$\begin{aligned}
\mathcal{L}'_\varepsilon(U) &= \{x \in \mathcal{L}'_\varepsilon \mid \overline{Q_\varepsilon(x)} \cap U \neq \emptyset\}, \\
\mathcal{L}_\varepsilon(U) &= \mathcal{L}'_\varepsilon(U) + \varepsilon\{z_1, \dots, z_{2^d}\}, \\
(\mathcal{L}'_\varepsilon(U))^\circ &= \{x \in \mathcal{L}'_\varepsilon \mid \overline{Q_\varepsilon(x)} \subset U\}, \\
(\mathcal{L}_\varepsilon(U))^\circ &= (\mathcal{L}'_\varepsilon(U))^\circ + \varepsilon\{z_1, \dots, z_{2^d}\}, \\
\partial\mathcal{L}'_\varepsilon(U) &= \mathcal{L}'_\varepsilon(U) \backslash (\mathcal{L}'_\varepsilon(U))^\circ, \\
\partial\mathcal{L}_\varepsilon(U) &= \partial\mathcal{L}'_\varepsilon(U) + \varepsilon\{z_1, \dots, z_{2^d}\}.
\end{aligned}$$

Furthermore let

$$U^\varepsilon = \bigcup_{\bar{x} \in \mathcal{L}'_\varepsilon(U)} Q_\varepsilon(\bar{x}), \quad U_\varepsilon = \bigcup_{\bar{x} \in (\mathcal{L}'_\varepsilon(U))^\circ} Q_\varepsilon(\bar{x}).$$

A lattice deformation should be thought of as a mapping $\mathcal{L}_\varepsilon \cap \Omega \to \mathbb{R}^d$. Choosing a suitable extension (e.g., by 0) and piecewise constant interpolation, we can and will assume that the lattice deformations $\mathcal{B}_\varepsilon(\Omega)$ are the functions $\Omega \to \mathbb{R}^d$, which are constant on every cell $Q_\varepsilon(x)$, $x \in \mathcal{L}'_\varepsilon(\Omega)$. (This will not change the energy, see below.)

If we have a deformation $y \in \mathcal{B}_\varepsilon(\Omega)$ and $x \in \Omega_\varepsilon$, we set $y_i(x) = y(\bar{x} + \varepsilon z_i)$,

$$\bar{y}(x) = \frac{1}{2^d} \sum_{i=1}^{2^d} y_i(x) \quad \text{and} \quad \bar{\nabla} y(x) = \frac{1}{\varepsilon}(y_1(x) - \bar{y}(x), \dots, y_{2^d}(x) - \bar{y}(x)).$$

Let $\mathcal{A}(U)$ be the set of all bounded open subsets of $U \subset \mathbb{R}^d$ and $\mathcal{A}_L(U)$ the set of all those, that have a Lipschitz boundary. In the following, we will consider a set $\Omega \in \mathcal{A}_L(\mathbb{R}^d)$ and the energies $F_\varepsilon \colon L^p(\Omega; \mathbb{R}^d) \times \mathcal{A}(\Omega) \to [0, \infty]$ for some fixed $1 < p < \infty$, defined by

$$F_\varepsilon(y, U) = \begin{cases} \varepsilon^d \displaystyle\sum_{x \in (\mathcal{L}'_\varepsilon(U))^\circ} W_{\text{cell}}(\bar\nabla y(x)) & \text{if } y \in \mathcal{B}_\varepsilon(U), \\ \infty & \text{otherwise.} \end{cases} \tag{1.1}$$

In this definition the energy only depends on the values of $y$ in $(\mathcal{L}_\varepsilon(U))^\circ \subset \mathcal{L}_\varepsilon \cap U$. Of course, there can be some points in $\mathcal{L}_\varepsilon \cap U$ which we do not use at all, but this is negligible if we impose Dirichlet boundary conditions as we will do later on.

We make some assumptions on the cell energy $W_{\text{cell}} \colon \mathbb{R}^{d \times 2^d} \to [0, \infty)$. Note, that a discrete gradient can take values precisely in the space

$$V_0 = \left\{ F \in \mathbb{R}^{d \times 2^d} : \sum_{j=1}^{2^d} a_{ij} = 0, \text{ for every } i = 1, \dots, d \right\}.$$

Therefore, we are only interested in the values of $W_{\text{cell}}$ on $V_0$.

**Assumption 1.** There are $c, c' > 0$ such that for every $F \in V_0$

$$W_{\text{cell}}(F) \geq c|F|^p - c'.$$

**Assumption 2.** There is a $c > 0$ such that for every $F \in V_0$

$$W_{\text{cell}}(F) \leq c(|F|^p + 1).$$

While these conditions are supposed to hold true for all our results, we also state a third assumption, which, if satisfied, allows for an application of the Cauchy-Born rule locally near $SO(d)$. The so-called Cauchy-Born energy density is defined by letting each atom follow the macroscopic gradient:

$$W_{\text{CB}}(M) := \frac{1}{|\det A|} W_{\text{cell}}(MZ)$$

for $M \in \mathbb{R}^{d \times d}$.

**Assumption 3.**     (i) $W_{\text{cell}} : \mathbb{R}^{d \times 2^d} \to \mathbb{R}$ is invariant under translations and rotations, i.e. for $F \in \mathbb{R}^{d \times 2^d}$,

$$W_{\text{cell}}(RF + (c, \dots, c)) = W_{\text{cell}}(F)$$

for all $R \in SO(d)$, $c \in \mathbb{R}^d$.

(ii) $W_{\text{cell}}(F)$ is minimal ($= 0$) if and only if there exists $R \in SO(d)$ and $c \in \mathbb{R}^d$ such that

$$F = RZ + (c, \dots, c).$$

(iii) $W_{\text{cell}}$ is $C^2$ in a neighborhood of $\bar{SO}(d) := SO(d)Z$ and the Hessian $D^2 W_{\text{cell}}(Z)$ at the identity is positive definite on the orthogonal complement of the subspace spanned by translations $(c, \dots, c)$ and infinitesimal rotations $FZ$, with $F^T = -F$.

(iv) $p \geq d$, which together with Assumption 1 implies in particular that $W_{\text{cell}}$ satisfies the growth assumption

$$\liminf_{\substack{|F| \to \infty \\ F \in V_0}} \frac{W_{\text{cell}}(F)}{|F|^d} > 0.$$

## Γ-convergence and Integral Representations

In our analysis, we consider energies on discrete systems depending on a small parameter $\varepsilon$, the scale of the lattice spacing. To make the limit for $\varepsilon \to 0$ precise and gain some knowledge about the behavior of associated minimizers, we will use De Giorgi's Γ-convergence. We recall the definition and some basic properties that will be needed in the sequel.

**Definition 1.12.** Let $X$ be a metric space and $F_n, F \colon X \to \bar{\mathbb{R}} = \mathbb{R} \cup \{-\infty, \infty\}$. We say $F_n$ Γ$(X)$-converges to $F$ ($F_n \xrightarrow{\Gamma} F$), if

(i) (liminf-inequality) For every $y, y_n \in X$ with $y_n \to y$, we have

$$F(y) \leq \liminf_{n \to \infty} F_n(y_n),$$

(ii) (recovery sequence) For every $y \in X$, there is a sequence $y_n \in X$ such that

$$F(y) \geq \limsup_{n \to \infty} F_n(y_n).$$

If $(F_\varepsilon)_{\varepsilon>0}$ is a family of functionals depending on a positive real parameter $\varepsilon$, we say $F_\varepsilon$ $\Gamma(X)$-converges to $F$, if for every sequence $\varepsilon_n > 0$ converging to 0, we have $F_{\varepsilon_n} \xrightarrow{\Gamma} F$. We will also use the $\Gamma$-lim sup and the $\Gamma$-lim inf, given by

$$F'(y) = \Gamma(X)\text{-}\liminf_{n\to\infty} F_n(y) = \inf\{\liminf_{n\to\infty} F_n(y_n)\colon y_n \to y \text{ in } X\},$$
$$F''(y) = \Gamma(X)\text{-}\limsup_{n\to\infty} F_n(y) = \inf\{\limsup_{n\to\infty} F_n(y_n)\colon y_n \to y \text{ in } X\}.$$

Note that (i) is equivalent to $F \le F'$ and (ii) is equivalent to $F \ge F''$. Hence, $F_n \xrightarrow{\Gamma} F$ if and only if $F' = F'' = F$. Furthermore, we see that $\Gamma$-convergence is a pointwise property, so we can speak about $\Gamma$-convergence at a specific point.

In the following proposition we assemble some basic properties of $\Gamma$-convergence that we will not prove here.

**Proposition 1.13.** *(i) The infima in the definitions of $F'$ and $F''$ are actually attained minima in $\mathbb{R}$;*

*(ii) every sequence of functionals on a separable metric space, like $L^p(U;\mathbb{R}^d)$, has a $\Gamma$-convergent subsequence;*

*(iii) $F'$, $F''$ and $F$ are lower semicontinuous with respect to convergence in $X$.*

*(iv) $\Gamma$-convergence satisfies the Urysohn property, i.e., $F_n$ $\Gamma$-converges to $F$, if and only if every subsequence of $F_n$ has a further subsequence, that $\Gamma$-converges to $F$;*

*(v) if $F_n$ $\Gamma$-converges to $F$ and $G_n$ converges uniformly on bounded sets to a continuous functional $G$, then $F_n + G_n$ $\Gamma$-converges to $F + G$.*

In view of applications, the most interesting property of $\Gamma$-convergence is the following theorem.

**Theorem 1.14.** *If $F_n$ $\Gamma$-converges to $F$ and sequences $y_n$ in $X$ with equibounded $F_n(y_n)$ are pre-compact then $F$ attains its minimum on $X$ and we have*

$$\min_{x\in X} F(x) = \lim_{n\to\infty} \inf_{x\in X} F_n(x).$$

*Furthermore, let $y_n \in X$ be a sequence with*

$$F_n(y_n) \to \lim_{n \to \infty} \inf_{x \in X} F_n(x),$$

*then the limit of every converging subsequence of $y_n$ is a minimizer of*
*$F$.*

For proofs of Proposition 1.13 and Theorem 1.14 see, e.g., [DM93].

Returning to our specific setting, for a sequence $\varepsilon_n > 0$ such that
$\varepsilon_n \to 0$, we define

$$\begin{aligned} F'(y, U) &:= \Gamma(L^p(\Omega; \mathbb{R}^d))\text{-} \liminf_{n \to \infty} F_{\varepsilon_n}(y, U) \\ &= \min\{\liminf_{n \to \infty} F_{\varepsilon_n}(y_n, U) \colon y_n \to y \text{ in } L^p(\Omega; \mathbb{R}^d)\}, \\ F''(y, U) &:= \Gamma(L^p(\Omega; \mathbb{R}^d))\text{-} \limsup_{n \to \infty} F_{\varepsilon_n}(y, U). \\ &= \min\{\limsup_{n \to \infty} F_{\varepsilon_n}(y_n, U) \colon y_n \to y \text{ in } L^p(\Omega; \mathbb{R}^d)\}. \end{aligned}$$

The limiting functionals we will encounter in the next section are
integral functionals of the form

$$I \colon W^{1,p}(U; \mathbb{R}^k) \to [0, \infty], \qquad I(y) = \int_U f(\nabla y(x)) \, dx$$

with $1 < p < \infty$, $U \in \mathcal{A}(\mathbb{R}^d)$, $f \colon \mathbb{R}^{k \times d} \to [0, \infty)$ continuous. Recall
that a Borel measurable and locally bounded function $f \colon \mathbb{R}^{k \times d} \to \mathbb{R}$ is
quasiconvex, if

$$f(M) \le \fint_U f(M + \nabla\varphi(x)) \, dx,$$

for every nonempty $U \in \mathcal{A}(\mathbb{R}^d)$, $M \in \mathbb{R}^{k \times d}$ and $\varphi \in W_0^{1,\infty}(U; \mathbb{R}^k)$. In
our analysis, the quasiconvexity of $f$ will be due to the following result.

**Theorem 1.15.** *If $I$ is sequentially weakly lower semicontinuous in*
*$W^{1,p}(U; \mathbb{R}^k)$, then $f$ is quasiconvex.*

A detailed discussion of quasiconvexity and related properties, including proofs of the above statements, is given, e.g., in [Dac08].

In order to guarantee that indeed our limiting functional is an integral functional, we will resort to the following general integral representation result on Sobolev spaces.

**Theorem 1.16.** *Let $1 \leq p < \infty$ and let $F \colon W^{1,p}(\Omega; \mathbb{R}^d) \times \mathcal{A}(\Omega) \to [0, \infty]$ satisfy the following conditions:*

  (i) *(locality) $F(y,U) = F(v,U)$, if $y(x) = v(x)$ for a.e. $x \in U$;*

  (ii) *(measure property) $F(y, \cdot)$ is the restriction of a Borel measure to $\mathcal{A}(\Omega)$;*

 (iii) *(growth condition) there exists $c > 0$ such that*

$$F(y,U) \leq c \int_U |\nabla y(x)|^p + 1 \, dx;$$

 (iv) *(translation invariance in y) $F(y,U) = F(y+a,U)$ for every $a \in \mathbb{R}^d$ ;*

  (v) *(lower semicontinuity) $F(\cdot, U)$ is sequentially lower semicontinuous with respect to weak convergence in $W^{1,p}(\Omega; \mathbb{R}^d)$;*

 (vi) *(translation invariance in x) With $y_M(x) = Mx$ we have*

$$F(y_M, B_r(x)) = F(y_M, B_r(x'))$$

*for every $M \in \mathbb{R}^{d \times d}$, $x, x' \in \Omega$ and $r > 0$ such that $B_r(x), B_r(x') \subset \Omega$.*

*Then there exists a continuous $f \colon \mathbb{R}^{d \times d} \to [0, \infty)$ such that*

$$0 \leq f(M) \leq C(1 + |M|^p) \text{ for every } M \in \mathbb{R}^{d \times d} \text{ and}$$
$$F(y,U) = \int_U f(\nabla y(x)) \, dx.$$

A proof can be found in [BD98, pp.77-81] or in the scalar-valued setting, which is essentially the same, in [DM93, pp.215-220].

To show the measure property in the previous theorem, we will use the following lemma.

**Lemma 1.17** (De Giorgi-Letta). *Let $X$ be a metric space with open sets $\tau$. Assume that $\rho \colon \tau \to [0, \infty]$ is an increasing set function such that*

  (i) $\rho(\emptyset) = 0$,

*(ii) (subadditivity) $\rho(U \cup V) \le \rho(U) + \rho(V)$ for all $U, V \in \tau$,*

*(iii) (inner regularity) $\rho(U) = \sup\{\rho(V) \colon V \in \tau, V \subset\subset U\}$ for all $U \in \tau$,*

*(iv) (superadditivity) $\rho(U \cup V) \ge \rho(U) + \rho(V)$ for all $U, V \in \tau$ with $U \cap V = \emptyset$.*

*Then the extension $\mu$ of $\rho$ to all subsets of $X$, defined by*

$$\mu(E) = \inf\{\rho(U) \colon U \in \tau, E \subset U\},$$

*is an outer measure and every Borel set is $\mu$-measurable.*

For a proof see, e.g., [FL07, pp.32-34].

## 1.3   A General Representation Result

In this section we will prove a general compactness and representation result for sequences of discrete deformations. For pair interactions, the following theorem has first been established by Alicandro and Cicalese in [AC04].

### Statement of the Representation Result

**Theorem 1.18** (compactness and integral representation). *Suppose Assumptions 1 and 2 are true. For every sequence $\varepsilon_n > 0$ such that $\varepsilon_n \to 0$, there exists a subsequence $\varepsilon_{n_k}$ and a functional $F \colon L^p(\Omega; \mathbb{R}^d) \times \mathcal{A}(\Omega) \to [0, \infty]$ such that for every $U \in \mathcal{A}(\Omega)$ and $y \in W^{1,p}(\Omega; \mathbb{R}^d)$ the functionals $F_{\varepsilon_{n_k}}(\cdot, U)$ $\Gamma(L^p(\Omega; \mathbb{R}^d))$-converge to $F(\cdot, U)$ at $y$. Furthermore there exists a quasiconvex function $f \colon \mathbb{R}^{d \times d} \to [0, \infty)$ satisfying*

$$c|M|^p - c' \le f(M) \le c'(|M|^p + 1)$$

*for some $c, c' > 0$ such that*

$$F(y, U) = \int_U f(\nabla y(x))\, dx \quad \text{if } y \in W^{1,p}(\Omega; \mathbb{R}^d).$$

*In addition, if $U \in \mathcal{A}_L(\Omega)$ (in particular, if $U = \Omega$), we have*

$$F(y,U) = \begin{cases} \displaystyle\int_U f(\nabla y(x))\, dx & \text{if } y|_U \in W^{1,p}(U; \mathbb{R}^d), \\ \infty & \text{otherwise,} \end{cases}$$

*and the functionals $F_{\varepsilon_{n_k}}(\cdot, U)$ Γ-converge to $F(\cdot, U)$.*

## Interpolation

We now define the continuous and piecewise affine interpolation $\tilde{y}$ of $y$, similar to [Sch09]. Note, that it is not important to choose this specific interpolation. But it is an easy and convenient choice.

First consider the cell $A\left[-\frac{1}{2}, \frac{1}{2}\right]^d$ and $y\colon A\left\{-\frac{1}{2}, \frac{1}{2}\right\}^d \to \mathbb{R}^d$. On every corner (0-dimensional face) of the cell just take $\tilde{y} = y$. Inductively, assume we already have chosen a simplicial decomposition on every $(k-1)$-dimensional face and have interpolated affine there. Let $F = \mathrm{co}\{z_{i_1}, \dots, z_{i_{2^k}}\}$ be a $k$-dimensional face. Set

$$\bar{z} = \frac{1}{2^k} \sum_{m=1}^{2^k} z_{i_m}, \quad \tilde{y}(\bar{z}) = \frac{1}{2^k} \sum_{m=1}^{2^k} y(z_{i_m}).$$

Now, decompose $F$ into the simplices $\mathrm{co}\{w_1, \dots, w_k, \bar{z}\}$, with simplices $\mathrm{co}\{w_1, \dots, w_k\}$ belonging to the simplicial decomposition of an $(n-1)$-dimensional face. Define $\tilde{y}$ to be the interpolation affine on every constructed simplex. If $y \in \mathcal{B}_\varepsilon(\Omega)$, we get $\tilde{y}$ on $\overline{\Omega_\varepsilon}$ by interpolating as above on every cell.

The following lemma is about the relation of $\bar{\nabla} y$ and $\nabla \tilde{y}$.

**Lemma 1.19.** *There are $C, c > 0$ such that for every $x \in \Omega_\varepsilon$ and $y \in \mathcal{B}_\varepsilon(\Omega)$*

$$c|\bar{\nabla} y(x)|^p \leq \fint_{Q_\varepsilon(x)} |\nabla \tilde{y}(x')|^p \, dx' \leq C|\bar{\nabla} y(x)|^p. \tag{1.2}$$

*Proof.* The statement is obviously independent of $\varepsilon, x$ and $\Omega$. We just have to consider the interpolation on one cell. But since

$$\|y|_Q\| = \left( \fint_Q |\nabla \tilde{y}(x')|^p \, dx' \right)^{\frac{1}{p}}$$

is a norm on $V_0$, the statement is a consequence of the equivalence of all norms on a finite dimensional space. $\square$

**Lemma 1.20.** *Let $\varepsilon_n > 0$, with $\varepsilon_n \to 0$, $y_n \in \mathcal{B}_{\varepsilon_n}(\Omega)$, and $y \in L^p_{loc}(\Omega;\mathbb{R}^d)$. Then $y_n \to y$ in $L^p_{loc}(\Omega;\mathbb{R}^d)$, if and only if $\tilde{y}_n \to y$ in $L^p_{loc}(\Omega;\mathbb{R}^d)$.*

*Proof.* First assume $y_n \to y$ in $L^p_{loc}(\Omega;\mathbb{R}^d)$. Let $V \subset\subset \Omega$ be arbitrary and $V \subset\subset W \subset\subset \Omega$. It is enough to show $\|y_n - \tilde{y}_n\|_{L^p(V;\mathbb{R}^d)} \to 0$. Let $\lambda_i \colon \mathbb{R}^d \to [0,1]$ denote the cell-periodic functions such that

$$\tilde{y}_n(x) = \sum_{i=1}^{2^d} \lambda_i\left(\frac{x}{\varepsilon_n}\right) y_n(\bar{x} + \varepsilon_n z_i)$$

$$= \sum_{i=1}^{2^d} \lambda_i\left(\frac{x}{\varepsilon_n}\right) y_n(x + \varepsilon_n(z_i - z_1)),$$

where without loss of generality we have chosen a numbering of $A\{-\frac{1}{2}, \frac{1}{2}\}^d$ such that $z_1 = A(-\frac{1}{2}, \ldots, -\frac{1}{2})$. Of course, $\lambda_i \geq 0$ and the $\lambda_i$ add up to 1 in any point and so for $n$ large enough

$$\int_V |y_n(x) - \tilde{y}_n(x)|^p \, dx$$

$$\leq \int_V \left(\sum_{i=1}^{2^d} \lambda_i\left(\frac{x}{\varepsilon_n}\right) |y_n(x) - y_n(x + \varepsilon_n(z_i - z_1))|\right)^p dx$$

$$\leq \int_V \max_{i=1,\ldots,2^d} |y_n(x) - y_n(x + \varepsilon_n(z_i - z_1))|^p \, dx$$

$$\leq \sum_{i=1}^{2^d} \int_V |y_n(x) - y_n(x + \varepsilon_n(z_i - z_1))|^p \, dx.$$

But the last term goes to 0 since for every $i \in \{1, \ldots, 2^d\}$

$$\|y_n - y_n(\cdot + \varepsilon_n(z_i - z_1))\|_{L^p(V;\mathbb{R}^d)}$$
$$\leq 2\|y_n - y\|_{L^p(W;\mathbb{R}^d)} + \|y - y(\cdot + \varepsilon_n(z_i - z_1))\|_{L^p(V;\mathbb{R}^d)}$$
$$\to 0.$$

Now, let $\tilde{y}_n \to y$ in $L^p_{loc}(\Omega; \mathbb{R}^d)$. Let $V \subset\subset \Omega$ be arbitrary and $V \subset\subset W \subset\subset \Omega$. Define $\max \tilde{y}_n$ and $\min \tilde{y}_n$ for each component on each simplex to be the max/min of that component of $\tilde{y}_n$ on that simplex. It follows from [AFG00, Prop. A.1], just substituting $L^1$ with $L^p$, that $\max \tilde{y}_n, \min \tilde{y}_n \to y$ in $L^p_{loc}(\Omega; \mathbb{R}^d)$. Since, $\tilde{y}_n$ takes the values $y_n(x)$ and $\tilde{y}_n(x)$ on the same cell, we can compare those values by using two simplices and get

$$\int_V |y(x) - y_n(x)|^p \, dx$$

$$\leq C\left( \int_V |y(x) - \tilde{y}_n(x)|^p \, dx + \int_W |\max \tilde{y}_n - \min \tilde{y}_n|^p \, dx \right)$$

$$\to 0.$$

$\square$

## Preliminary Lemmata

We proceed to collect further lemmata. We will use them later to verify the requirements of Theorem 1.16. In the following, fix some sequence of positive real numbers $\varepsilon_n \to 0$.

**Lemma 1.21.** *Suppose Assumption 1 is true. If $y \in L^p(\Omega; \mathbb{R}^d)$ and $U \in \mathcal{A}(\Omega)$ are such that $F'(y, U) < \infty$, then $y \in W^{1,p}(U; \mathbb{R}^d)$ and*

$$F'(y, U) \geq c\|\nabla y\|^p_{L^p(U; \mathbb{R}^{d \times d})} - c'|U|, \tag{1.3}$$

*for some $c, c' > 0$ independent of $y$ and $U$.*

*Proof.* Let $y_n \to y$ in $L^p(\Omega; \mathbb{R}^d)$ such that $\liminf_{n\to\infty} F_{\varepsilon_n}(y_n, U) < \infty$. For some subsequence $n_k$, we have

$$\lim_{k\to\infty} F_{\varepsilon_{n_k}}(y_{n_k}, U) = \liminf_{n\to\infty} F_{\varepsilon_n}(y_n, U),$$

$y_{n_k} \in \mathcal{B}_{\varepsilon_{n_k}}(U)$ and $F_{\varepsilon_{n_k}}(y_{n_k}, U) \leq M < \infty$ for some fixed $M > 0$. By Lemma 1.20 we have $\tilde{y}_{n_k} \to y$ in $L^p(V, \mathbb{R}^d)$ for every $V \subset\subset U$.

Furthermore, by Assumption 1 and Lemma 1.19, we get

$$M \geq F_{\varepsilon_{n_k}}(y_{n_k}, U) = \varepsilon_{n_k}^d \sum_{x \in (\mathcal{L}'_{\varepsilon_{n_k}}(U))^\circ} W_{\text{cell}}(\bar{\nabla} y_{n_k}(x))$$

$$\geq \varepsilon_{n_k}^d \sum_{x \in (\mathcal{L}'_{\varepsilon_{n_k}}(U))^\circ} \left( c|\bar{\nabla} y_{n_k}(x)|^p - c' \right)$$

$$\geq \varepsilon_{n_k}^d \sum_{x \in (\mathcal{L}'_{\varepsilon_{n_k}}(U))^\circ} \left( c \fint_{Q_{\varepsilon_{n_k}}(x)} |\nabla \tilde{y}_{n_k}(x')|^p \, dx' - c' \right).$$

We thus obtain

$$c \int_{U_{\varepsilon_{n_k}}} |\nabla \tilde{y}_{n_k}(x')|^p \, dx' \leq M + c'|U|,$$

hence the gradients are bounded in $L^p(V; \mathbb{R}^d)$. Therefore $y \in W^{1,p}(V, \mathbb{R}^d)$ and $\nabla \tilde{y}_{n_k} \rightharpoonup \nabla y$ in $L^p(V; \mathbb{R}^d)$. Weak sequentially lower semicontinuity of the norm yields

$$c\|\nabla y\|_{L^p(V;\mathbb{R}^{d \times d})}^p \leq \liminf_{n \to \infty} F_{\varepsilon_n}(y_n, U) + c'|U|,$$

but the right hand side is independent of $V$, thus $y \in W^{1,p}(U, \mathbb{R}^d)$ and

$$c\|\nabla y\|_{L^p(U;\mathbb{R}^{d \times d})}^p \leq \liminf_{n \to \infty} F_{\varepsilon_n}(y_n, U) + c'|U|.$$

The definition of the $\Gamma$-lim inf now yields the lemma. $\qquad \square$

**Lemma 1.22.** *Suppose Assumption 2 is true. Then there is a $C > 0$ such that for every $V \in \mathcal{A}_L(\Omega)$, $U \in \mathcal{A}(V)$ and $y \in L^p(\Omega; \mathbb{R}^d) \cap W^{1,p}(V; \mathbb{R}^d)$ we have*

$$F''(y, U) \leq C \left( \|\nabla y\|_{L^p(U;\mathbb{R}^{d \times d})}^p + |U| \right). \tag{1.4}$$

*Proof.* We first prove (1.4) for every $y \in C_c^\infty(\mathbb{R}^d; \mathbb{R}^d)$. For $x \in \mathcal{L}'_{\varepsilon_n}$ and $a \in Q_{\varepsilon_n}(x)$ define

$$y_n(a) = y(x).$$

Thus $y_n \in \mathcal{B}_{\varepsilon_n}(U)$ and since $y$ is uniformly continuous, we have $y_n \to y$ uniformly and hence in $L^p(\Omega; \mathbb{R}^d)$. By Taylor expansion we have

$$\left| \frac{y_n(\bar{x}) - y_n(\bar{x} + \varepsilon_n z_i)}{\varepsilon_n} \right| \leq C(|\nabla y(\bar{x})| + \varepsilon_n \|\nabla^2 y\|_\infty).$$

With Assumption 2 we can calculate

$$
\begin{aligned}
F_{\varepsilon_n}(y_n, U) &= \varepsilon_n^d \sum_{x \in (\mathcal{L}'_{\varepsilon_n}(U))^\circ} W_{\text{cell}}(\bar{\nabla} y_n(x)) \\
&\le C\varepsilon_n^d \sum_{x \in (\mathcal{L}'_{\varepsilon_n}(U))^\circ} \left( |\bar{\nabla} y_n(x)|^p + 1 \right) \\
&\le C'|U| + C\varepsilon_n^d \sum_{x \in (\mathcal{L}'_{\varepsilon_n}(U))^\circ} \left( |\nabla y(x)|^p + \varepsilon_n^p \|\nabla^2 y\|_\infty^p \right) \\
&\le C'|U| + C''|U|\varepsilon_n^p \|\nabla^2 y\|_\infty^p + C\varepsilon_n^d \sum_{x \in (\mathcal{L}'_{\varepsilon_n}(U))^\circ} |\nabla y(x)|^p.
\end{aligned}
$$

Furthermore for every $x' \in Q_{\varepsilon_n}(x)$, $x \in (\mathcal{L}'_{\varepsilon_n}(U))^\circ$

$$
\begin{aligned}
|\nabla y(x)|^p &\le C(|\nabla y(x')|^p + |\nabla y(x) - \nabla y(x')|^p) \\
&\le C(|\nabla y(x')|^p + \varepsilon_n^p \|\nabla^2 y\|_\infty^p),
\end{aligned}
$$

and, by integrating over $x'$ and summing over $x$, we get

$$
\begin{aligned}
\varepsilon_n^d \sum_{x \in (\mathcal{L}'_{\varepsilon_n}(U))^\circ} |\nabla y(x)|^p &\le C \left( \int_{U_{\varepsilon_n}} |\nabla y(x')|^p \, dx' + |U_{\varepsilon_n}| \varepsilon_n^p \|\nabla^2 y\|_\infty^p \right) \\
&\le C \left( \int_U |\nabla y(x')|^p \, dx' + |U| \varepsilon_n^p \|\nabla^2 y\|_\infty^p \right).
\end{aligned}
$$

Putting the two inequalities together and letting $n \to \infty$, we obtain

$$
\limsup_{n \to \infty} F_{\varepsilon_n}(y_n, U) \le C \left( \|\nabla y\|_{L^p(U;\mathbb{R}^{d \times d})}^p + |U| \right).
$$

So by the definition of the Γ-lim sup we have (1.4).

The general case follows easily: Since $V$ has Lipschitz boundary, we can take $y_k \in C_c^\infty(\mathbb{R}^d; \mathbb{R}^d)$ such that $y_k \to y$ in $W^{1,p}(V; \mathbb{R}^d)$. Then we have by lower semicontinuity of $F''(\cdot, U)$

$$
\begin{aligned}
F''(y, U) &\le \liminf_{k \to \infty} F''(y_k, U) \\
&\le \liminf_{k \to \infty} C \left( \|\nabla y_k\|_{L^p(U;\mathbb{R}^{d \times d})}^p + |U| \right) \\
&= C \left( \|\nabla y\|_{L^p(U;\mathbb{R}^{d \times d})}^p + |U| \right).
\end{aligned}
$$

□

**Lemma 1.23.** *Suppose Assumptions 1 and 2 are true. Let $U, V, U' \in \mathcal{A}(\Omega)$ be such that $U' \subset\subset U$. Then for every $y \in W^{1,p}(\Omega; \mathbb{R}^d)$*

$$F''(y, U' \cup V) \leq F''(y, U) + F''(y, V).$$

*Proof.* Without loss of generality, we can assume the terms on the right hand side to be finite. According to the properties of the $\Gamma$-lim sup it is possible to find sequences $u_n, v_n$ such that

$$\limsup_{n \to \infty} F_{\varepsilon_n}(u_n, U) = F''(y, U), \qquad u_n \to y \text{ in } L^p(\Omega; \mathbb{R}^d),$$

$$\limsup_{n \to \infty} F_{\varepsilon_n}(v_n, V) = F''(y, V), \qquad v_n \to y \text{ in } L^p(\Omega; \mathbb{R}^d).$$

For $n$ large enough $F_{\varepsilon_n}(u_n, U)$ and $F_{\varepsilon_n}(v_n, V)$ are bounded and $u_n \in \mathcal{B}_{\varepsilon_n}(U)$, $v_n \in \mathcal{B}_{\varepsilon_n}(V)$.

Fix $N \in \mathbb{N}$, $N \geq 5$ and then define $D = \text{dist}(U', U^c)$ and

$$U_j = \left\{ x \in U : \text{dist}(x, U') < \frac{jD}{N} \right\}.$$

Choose cut-off functions $\varphi_j$ such that

$$\varphi_j(x) = 1 \quad \forall x \in U_j,$$
$$\varphi_j \in C_c^\infty(U_{j+1}; [0, 1]),$$
$$\|\nabla \varphi_j\|_\infty \leq \frac{2N}{D}.$$

Next we define

$$w_{n,j}(x) = \varphi_j(\bar{x}) u_n(x) + (1 - \varphi_j(\bar{x})) v_n(x)$$

and calculate

$$\frac{w_{n,j}(x + \varepsilon_n z_i) - w_{n,j}(x)}{\varepsilon_n}$$
$$= \varphi_j(\overline{x + \varepsilon_n z_i}) \frac{u_n(x + \varepsilon_n z_i) - u_n(x)}{\varepsilon_n}$$
$$+ (1 - \varphi_j(\overline{x + \varepsilon_n z_i})) \frac{v_n(x + \varepsilon_n z_i) - v_n(x)}{\varepsilon_n}$$
$$+ (u_n(x) - v_n(x)) \frac{\varphi_j(\overline{x + \varepsilon_n z_i}) - \varphi_j(\bar{x})}{\varepsilon_n}. \tag{1.5}$$

To estimate $F_{\varepsilon_n}(w_{n,j}, U' \cup V)$, we have to look at $(\mathcal{L}'_{\varepsilon_n}(U' \cup V))^\circ$. Clearly, if $x \in (\mathcal{L}'_{\varepsilon_n}(U_j))^\circ$, then $\bar{\nabla} w_{n,j}(x) = \bar{\nabla} u_n(x)$ and if $x \in (\mathcal{L}'_{\varepsilon_n}(V \backslash \overline{U_{j+1}}))^\circ$, then $\bar{\nabla} w_{n,j}(x) = \bar{\nabla} v_n(x)$. To control the other cases, observe that for $n$ large enough $\mathrm{diam}(Q_{\varepsilon_n}) \leq \frac{D}{2N}$ and thus

$$(\mathcal{L}'_{\varepsilon_n}(U' \cup V))^\circ \subset (\mathcal{L}'_{\varepsilon_n}(U_j))^\circ \cup (\mathcal{L}'_{\varepsilon_n}(V \backslash \overline{U_{j+1}}))^\circ \cup (\mathcal{L}'_{\varepsilon_n}(V \cap (U_{j+2} \backslash \overline{U_{j-1}})))^\circ$$

for every $j \in \{2, \ldots, N-3\}$ and $n$ large enough. With $W_j = V \cap (U_{j+2} \backslash \overline{U_{j-1}})$, we then have

$$F_{\varepsilon_n}(w_{n,j}, U' \cup V) = \varepsilon_n^d \sum_{x \in (\mathcal{L}'_{\varepsilon_n}(U' \cup V))^\circ} W_{\mathrm{cell}}(\bar{\nabla} w_{n,j}(x))$$

$$\leq F_{\varepsilon_n}(u_n, U) + F_{\varepsilon_n}(v_n, V) + \underbrace{\varepsilon_n^d \sum_{x \in (\mathcal{L}'_{\varepsilon_n}(W_j))^\circ} W_{\mathrm{cell}}(\bar{\nabla} w_{n,j}(x))}_{:=S_{j,n}}.$$

We now have to estimate $S_{j,n}$. For all $n$ large enough, use first Assumption 2 and then (1.5) to get

$$S_{j,n} \leq C \varepsilon_n^d \sum_{x \in (\mathcal{L}'_{\varepsilon_n}(W_j))^\circ} (|\bar{\nabla} w_{n,j}(x)|^p + 1)$$

$$\leq C \varepsilon_n^d \sum_{x \in (\mathcal{L}'_{\varepsilon_n}(W_j))^\circ} (|\bar{\nabla} u_n(x)|^p + |\bar{\nabla} v_n(x)|^p$$

$$+ |u_n(x) - v_n(x)|^p \|\nabla \varphi_j\|_\infty^p + 1)$$

$$\leq C \varepsilon_n^d \sum_{x \in (\mathcal{L}'_{\varepsilon_n}(W_j))^\circ} (|\bar{\nabla} u_n(x)|^p + |\bar{\nabla} v_n(x)|^p + |u_n(x) - v_n(x)|^p N^p + 1)$$

$$\leq C \int_{(W_j)_{\varepsilon_n}} |\nabla \tilde{u}_n(x)|^p + |\nabla \tilde{v}_n(x)|^p + N^p |u_n(x) - v_n(x)|^p + 1 \, dx,$$

because of the gradient of $\varphi$ being bounded by $CN$ and Lemma 1.19. Averaging over $j$, we get

$$\frac{1}{N-4} \sum_{j=2}^{N-3} S_{j,n} \leq C \frac{1}{N-4} \int_{V_{\varepsilon_n}} |\nabla \tilde{u}_n(x)|^p + |\nabla \tilde{v}_n(x)|^p + 1 \, dx$$

$$+ N^p \int_{V_{\varepsilon_n}} |u_n(x) - v_n(x)|^p \, dx. \qquad (1.6)$$

Of course we can always find a number $j(n)$ such that

$$S_{j(n),n} \leq \frac{1}{N-4} \sum_{j=2}^{N-3} S_{j,n}.$$

By Lemma 1.19 and Assumption 1, the first integral in (1.6) is bounded, but

$$\|u_n - v_n\|_{L^p(\Omega;\mathbb{R}^d)} \to 0$$

for $n \to \infty$, hence

$$\limsup_{n\to\infty} S_{j(n),n} \leq \frac{C}{N-4}.$$

If we define $y_n = w_{n,j(n)}$, then obviously $y_n \in \mathcal{B}_{\varepsilon_n}(U' \cup V)$ and $y_n \to y$ in $L^p(\Omega;\mathbb{R}^d)$. We have

$$F''(y,U' \cup V) \leq \limsup_{n\to\infty} F_{\varepsilon_n}(y_n, U' \cup V)$$
$$\leq \limsup_{n\to\infty} F_{\varepsilon_n}(u_n, U) + \limsup_{n\to\infty} F_{\varepsilon_n}(v_n, V) + \limsup_{n\to\infty} S_{j(n),n}$$
$$\leq F''(y,U) + F''(y,V) + \frac{C}{N-4}.$$

Letting $N \to \infty$, we get the conclusion. $\qquad\square$

**Lemma 1.24.** *Suppose Assumptions 1 and 2 are true. Then for every $V \in \mathcal{A}_L(\Omega)$, $U \in \mathcal{A}(V)$, and $y \in L^p(\Omega;\mathbb{R}^d) \cap W^{1,p}(V;\mathbb{R}^d)$*

$$F''(y,U) = \sup_{U' \subset\subset U} F''(y,U').$$

*Proof.* Since $F''(y,\cdot)$ is an increasing set function, we only have to show '$\leq$'.

Let $\delta > 0$. Then take a $U''' \subset\subset U$ such that

$$|U \backslash \overline{U'''}| + \|\nabla y\|_{L^p(U\backslash\overline{U'''};\mathbb{R}^d)} \leq \delta.$$

Choosing $U', U''$ such that

$$U''' \subset\subset U'' \subset\subset U' \subset\subset U,$$

we can calculate

$$F''(y,U) \leq F''(y,U'' \cup U\backslash\overline{U'''})$$
$$\leq F''(y,U') + F''(y,U\backslash\overline{U'''})$$
$$\leq F''(y,U') + \delta C,$$

where we used Lemma 1.23 and Lemma 1.22. $\qquad\square$

**Lemma 1.25.** *Suppose Assumptions 1 and 2 are true. Then for every $V \in \mathcal{A}_L(\Omega)$, $U \in \mathcal{A}(V)$, and $u, v \in L^p(\Omega; \mathbb{R}^d) \cap W^{1,p}(V; \mathbb{R}^d)$ such that $u(x) = v(x)$ for almost every $x \in U$, we have*

$$F''(u, U) = F''(v, U).$$

*Proof.* If $u = v$ a.e. in $U$ then for $U' \subset\subset U$ we have $F''(u, U') = F''(v, U')$. To see this, just change any approximating discrete sequence of $u$ outside of $(U')^{\varepsilon_n}$ such that the new sequence converges to $v$. But this is enough by Lemma 1.24. □

## Proof of the Representation Result

Now, we can finally prove the compactness result:

*Proof of Theorem 1.18.* First we find by a suitable diagonal argument a subsequence $F_{\varepsilon_{n_k}}$ such that we get Γ-convergence for every $U \in \mathcal{A}(\Omega)$. For this we define

$$\mathcal{A}_1 = \left\{ U \subset \Omega \colon U = \bigcup_{i=1}^{N} B_{r_i}(x_i), x_i \in \mathbb{Q}^d, r_i \in \mathbb{Q}, r_i > 0, N \in \mathbb{N}. \right\}$$

The set $\mathcal{A}_1$ is countable and we can write $\mathcal{A}_1 = \{U_1, U_2, \dots\}$. Now choose subsequences as follows:

$$
\begin{aligned}
F_{\varepsilon_n}(\cdot, U_1) \quad &\text{has a Γ-convergent subsequence} \quad F_{\varepsilon_{n_k^1}}(\cdot, U_1), \\
F_{\varepsilon_{n_k^1}}(\cdot, U_2) \quad &\text{has a Γ-convergent subsequence} \quad F_{\varepsilon_{n_k^2}}(\cdot, U_2), \\
F_{\varepsilon_{n_k^2}}(\cdot, U_3) \quad &\text{has a Γ-convergent subsequence} \quad F_{\varepsilon_{n_k^3}}(\cdot, U_3), \\
\vdots \qquad\qquad &\qquad\qquad\qquad \vdots \qquad\qquad\qquad\qquad \vdots
\end{aligned}
$$

Now setting $n_k = n_k^k$, we see that $F_{\varepsilon_{n_k}}(\cdot, U)$ Γ-converges to a $F(\cdot, U)$ for every $U \in \mathcal{A}_1$. In the following we will only consider the sequence $\varepsilon_{n_k}$ and, in particular, define $F'$ and $F''$ accordingly. Furthermore, we define $F(y, U) := F'(y, U)$ for every $y$ and $U$.

For $W \subset\subset U \subset \Omega$, by compactness of $\overline{W}$, we always find $V \in \mathcal{A}_1$ such that $W \subset V \subset\subset U$. Hence, by Lemma 1.24 we have

$$F''(y, U) = \sup\{F''(y, V) \colon V \subset\subset U, V \in \mathcal{A}_1\}$$

for every $U \in \mathcal{A}(\Omega)$ and $y \in W^{1,p}(\Omega, \mathbb{R}^d)$. Using, that $F'(y, \cdot)$ is an increasing set function, we can calculate

$$
\begin{aligned}
\sup\{F'(y, V) \colon V \subset\subset U, V \in \mathcal{A}_1\} &\leq F'(y, U) \\
&\leq F''(y, U) \\
&= \sup\{F''(y, V) \colon V \subset\subset U, V \in \mathcal{A}_1\}.
\end{aligned}
$$

But the first and the last term are equal, thus $F'(y, U) = F''(y, U) = F(y, U)$, whenever $y \in W^{1,p}(\Omega, \mathbb{R}^d)$.

The next step is to get an integral representation by showing that $F$, restricted to $W^{1,p}(\Omega; \mathbb{R}^d)$, satisfies the conditions (i)-(vi) in Theorem 1.16. We immediately see the locality (i), by Lemma 1.25, and the growth condition (iii), by Lemma 1.22. Furthermore, since the $F_{\varepsilon_{n_k}}$ are translation invariant in $y$, so is $F$, which yields (iv). To get the lower semicontinuity (v), just remember that weak convergence in $W^{1,p}(\Omega, \mathbb{R}^d)$ implies strong convergence in $L^p(\Omega, \mathbb{R}^d)$ and that $\Gamma(X)$-limits are sequentially lower semicontinuous with respect to the convergence in $X$.

To get the measure property (ii), it is enough to show that we can apply the De-Giorgi-Letta criterion (Lemma 1.17) with $\rho = F(y, \cdot)$. Obviously $F(y, \cdot)$ is an increasing set function and $F(y, \emptyset) = 0$. Remark that for every $W \subset\subset U \cup V (W, U, V$ open), there are open sets $U', V'$ such that $U' \subset\subset U$, $V' \subset\subset V$ and $W \subset U' \cup V'$, which is easily seen by the compactness of $\overline{W}$. Hence the subadditivity follows from the Lemmata 1.23 and 1.24. The inner regularity is explicitly given by Lemma 1.24. The superadditivity we can show directly. Take a sequence $y_k \in \mathcal{B}_{\varepsilon_{n_k}}(U \cup V)$ such that $y_k \to y$ in $L^p(\Omega; \mathbb{R}^d)$ and

$$
F(y, U \cup V) = \lim_{k \to \infty} F_{\varepsilon_{n_k}}(y_k, U \cup V).
$$

Then,

$$
\begin{aligned}
F(y, U \cup V) &\geq \liminf_{k \to \infty} F_{\varepsilon_{n_k}}(y_k, U) + \liminf_{k \to \infty} F_{\varepsilon_{n_k}}(y_k, V) \\
&\geq F(y, U) + F(y, V),
\end{aligned}
$$

since $U \cap V = \emptyset$. Hence, we can apply the De-Giorgi-Letta criterion and obtain (ii). Finally, condition (vi) states that for every $M \in \mathbb{R}^{d \times d}$, $z, z' \in \Omega$ and $r > 0$ such that $B_r(z), B_r(z') \subset \Omega$, we have

$$
F(y_M, B_r(z)) = F(y_M, B_r(z')),
$$

if we set $y_M(x) = Mx$. By inner regularity, it is enough to show that, for any $r' < r$,

$$F(y_M, B_r(z)) \geq F(y_M, B_{r'}(z')).$$

Let $y_k \in \mathcal{B}_{\varepsilon_{n_k}}(B_r(z))$ such that $y_k \to y_M$ in $L^p(\Omega; \mathbb{R}^d)$ and

$$\lim_{k \to \infty} F_{\varepsilon_{n_k}}(y_k, B_r(z)) = F(y_M, B_r(z)).$$

Denote by $a_k$ the only point in $\mathcal{L}_{\varepsilon_{n_k}} \cap Q_{\varepsilon_{n_k}}(z' - z)$. Then define

$$u_k(x) = \begin{cases} y_k(x - a_k) + Ma_k & \text{if } x \in (B_{r'}(z'))^{\varepsilon_{n_k}} \\ M\bar{x} & \text{else.} \end{cases}$$

If $k$ is large enough, then $x - a_k \in (B_r(z))_{\varepsilon_{n_k}}$, $u_k \in \mathcal{B}_{\varepsilon_{n_k}}(B_{r'}(z'))$ and

$$\bar{\nabla} u_k(x) = \bar{\nabla} y_k(x - a_k)$$

for all $x \in (B_{r'}(z'))^{\varepsilon_{n_k}}$. Hence,

$$F_{\varepsilon_{n_k}}(u_k, B_{r'}(z')) \leq F_{\varepsilon_{n_k}}(y_k, B_r(z)).$$

Furthermore, we have $Ma_k \to M(z' - z)$ and $y_k(\cdot - a_k) \to M(\cdot - (z' - z))$ in $L^p(B_{r'}(z'); \mathbb{R}^d)$ and therefore $u_k \to y_M$ in $L^p(\Omega; \mathbb{R}^d)$. Hence, we get

$$\begin{aligned} F(y_M, B_{r'}(z')) &\leq \liminf_{k \to \infty} F_{\varepsilon_{n_k}}(u_k, B_{r'}(z')) \\ &\leq \liminf_{k \to \infty} F_{\varepsilon_{n_k}}(y_k, B_r(z)) \\ &= F(y_M, B_r(z)), \end{aligned}$$

and (vi) is proven.

Consequently, we can apply Theorem 1.16 to the restriction of $F$ to $W^{1,p}(\Omega; \mathbb{R}^d) \times \mathcal{A}(\Omega)$. In particular, there is a continuous function $f \colon \mathbb{R}^{d \times d} \to [0, \infty)$ such that

$$F(y, U) = \int_U f(\nabla y(x)) \, dx \quad \text{if } y \in W^{1,p}(\Omega; \mathbb{R}^d)$$

and

$$0 \leq f(M) \leq C(1 + |M|^p) \text{ for every } M \in \mathbb{R}^{d \times d}. \tag{1.7}$$

The asserted lower bound on $f$ is instantly obtained, if we apply Lemma 1.21 to $y_M$ and use the integral representation. And finally, $f$ is quasiconvex by Theorem 1.15, since $F(\cdot, \Omega)$ is sequentially lower semicontinuous with respect to weak convergence in $W^{1,p}(\Omega; \mathbb{R}^d)$.

For $U$ with Lipschitz boundary take $y \in L^p(\Omega; \mathbb{R}^d) \cap W^{1,p}(U, \mathbb{R}^d)$. By Lemma 1.24, we have

$$F''(y, U) = \sup\{F''(y, V) \colon V \subset\subset U, V \in \mathcal{A}_1\}.$$

Using that $F'(y, \cdot)$ is an increasing set function, we can calculate

$$\begin{aligned} \sup\{F'(y, V) \colon V \subset\subset U, V \in \mathcal{A}_1\} &\leq F'(y, U) \\ &\leq F''(y, U) \\ &= \sup\{F''(y, V) \colon V \subset\subset U, V \in \mathcal{A}_1\}. \end{aligned}$$

But the first and the last term are equal, thus $F'(y, U) = F''(y, U) = F(y, U)$. If $y \in L^p(\Omega; \mathbb{R}^d) \backslash W^{1,p}(U, \mathbb{R}^d)$, then

$$\infty = F'(y, U) = F''(y, U) = F(y, U)$$

by Lemma 1.21. Hence, $F_{\varepsilon_{n_k}}(\cdot, U)$ $\Gamma(L^p(\Omega; \mathbb{R}^d))$-converges to $F(\cdot, U)$. To get the integral representation for $y \in L^p(\Omega; \mathbb{R}^d) \cap W^{1,p}(U, \mathbb{R}^d)$, observe, that since $U$ has Lipschitz boundary, we can find a function $v \in W^{1,p}(\Omega; \mathbb{R}^d)$ such that

$$y(x) = v(x) \quad \text{for almost every } x \in U.$$

Then, by Lemma 1.25,

$$F(y, U) = F(v, U) = \int_U f(\nabla v(x)) \, dx = \int_U f(\nabla y(x)) \, dx.$$

$\square$

## 1.4  The Boundary Value Problem

While loading terms can be included in our results so far without difficulties, the restriction to deformations with preassigned boundary values is more subtle.

### Statement of Representation Result with Boundary Conditions

Suppose $g \in W^{1,\infty}(\mathbb{R}^d; \mathbb{R}^d)$ is a boundary datum. We will then always choose the precise representative for $g$ and thus assume that $g$ is continuous. We define the admissible lattice deformations $\mathcal{B}_\varepsilon(U, g)$ as the functions in $\mathcal{B}_\varepsilon(U)$, that satisfy the boundary condition

$$y(x) = g(\bar{x}), \text{ whenever } x \in \partial \mathcal{L}_\varepsilon(U).$$

The correspondingly restricted discrete functional is

$$F_\varepsilon^g(y, U) = \begin{cases} F_\varepsilon(y, U) & \text{if } y \in \mathcal{B}_\varepsilon(U, g), \\ \infty & \text{otherwise.} \end{cases}$$

Assume that $\varepsilon_{n_k}$ and $f$ are as in Theorem 1.18, let us for simplicity write just $\varepsilon_k$ in the following and set

$$F^g(y, U) = \begin{cases} F(y, U) & \text{if } y|_U \in g + W_0^{1,p}(U; \mathbb{R}^d), \\ \infty & \text{otherwise.} \end{cases}$$

In analogy to Theorem 1.18 we then have:

**Theorem 1.26.** *Suppose Assumptions 1 and 2 are true, we have* $g \in W^{1,\infty}(\mathbb{R}^d; \mathbb{R}^d)$*, and* $F^g$*,* $F_{\varepsilon_k}^g$ *are as above. Then* $F_{\varepsilon_k}^g(\cdot, U)$ $\Gamma(L^p(\Omega; \mathbb{R}^d))$*-converges to* $F^g(\cdot, U)$ *for every* $U \in \mathcal{A}_L(\Omega)$*.*

## Improved results on Interpolations

We start by improving Lemma 1.20 for sequences in $\mathcal{B}_\varepsilon(U, g)$. This is possible, because now we can control what happens near the boundary. Note, that now we can naturally define the interpolation $\tilde{y}$ on all of $U$, namely, we just extend $y$ by the discretization of $g$ before we interpolate.

**Lemma 1.27.** *Let* $U \in \mathcal{A}_L(\Omega)$ *and* $y_k \in \mathcal{B}_{\varepsilon_k}(U, g)$*. Then* $y_k \to y$ *in* $L^p(U; \mathbb{R}^d)$ *if and only if* $\tilde{y}_k \to y$ *in* $L^p(U; \mathbb{R}^d)$*.*

*Proof.* Choose some open bounded set $U'$ with Lipschitz boundary and $U \subset\subset U'$. Extend the functions by defining $y_k(x) := g(\bar{x})$ and $y(x) := g(x)$ for $x \in U' \backslash U$. So $y_k \in \mathcal{B}_{\varepsilon_k}(U', g)$ and, since $g$ is Lipschitz, we have that $y_k \to y$ in $L^p(U; \mathbb{R}^d)$ implies $y_k \to y$ in $L^p(U'; \mathbb{R}^d)$ and $\tilde{y}_k \to y$ in $L^p(U; \mathbb{R}^d)$ implies $\tilde{y}_k \to y$ in $L^p(U'; \mathbb{R}^d)$. But then the statement follows from Lemma 1.20. $\qquad\square$

## Proof of the Boundary Value Representation Result

*Proof of Theorem 1.26.* Fix $U \in \mathcal{A}_L(\Omega)$. We start with the lim inf-inequality. Let $y_k, y \in L^p(\Omega; \mathbb{R}^d)$ such that $y_k \to y$. We can assume that

$$\liminf_{k\to\infty} F_{\varepsilon_k}^g(y_k, U) < \infty,$$

because otherwise there is nothing to show. For some subsequence we then get

$$\liminf_{k\to\infty} F^g_{\varepsilon_k}(y_k, U) = \lim_{l\to\infty} F^g_{\varepsilon_{k_l}}(y_{k_l}, U).$$

But since $F_{\varepsilon_{k_l}} \le F^g_{\varepsilon_{k_l}}$, we can argue as in Lemma 1.21 to see that $y \in W^{1,p}(U; \mathbb{R}^d)$ and, for any $V \subset\subset U$, that $\tilde{y}_{k_l} \rightharpoonup y$ in $W^{1,p}(V; \mathbb{R}^d)$. Using Lemma 1.27, we see that $\tilde{y}_{k_l}$ converges strongly in $L^p(U; \mathbb{R}^d)$ and, since $\nabla \tilde{y}_{k_l}$ is now bounded in $L^p(U; \mathbb{R}^d)$, weakly in $W^{1,p}(U; \mathbb{R}^d)$ to $y$. Regarding the boundary condition, there are open neighborhoods $V_l$ of $\partial U$, where $\tilde{y}_{k_l}$ is an affine interpolation of $g$. Namely, $V_l$ is the interior of the union of all cells $Q_{\varepsilon_{k_l}}$, with $\overline{Q_{\varepsilon_{k_l}}} \cap \partial U \neq \emptyset$. Then

$$\sup_{x\in\partial U} |\tilde{y}_{k_l}(x) - g(x)| \le \sup_{x\in V_l} |\tilde{y}_{k_l}(x) - g(x)| \le C\varepsilon_{k_l}$$

since $g$ is Lipschitz. Denoting the trace operator by $T$, we thus have $T\tilde{y}_{k_l} \rightharpoonup Ty = Tg$ in $L^p(\partial U; \mathbb{R}^d)$ and hence $y \in g + W^{1,p}_0(U; \mathbb{R}^d)$. But then, we can calculate

$$F^g(y, U) = F(y, U) \le \liminf_{k\to\infty} F_{\varepsilon_k}(y_k, U) \le \liminf_{k\to\infty} F^g_{\varepsilon_k}(y_k, U),$$

and have indeed proven the $\liminf$-inequality.

To get the $\Gamma$-convergence result, we now proof the $\limsup$-inequality. Let us first assume $y(x) = g(x) + \psi(x)$, for every $x \in U$ and some $\psi \in C^\infty_c(U; \mathbb{R}^d)$. Then

$$F^g(y, U) = F(y, U) < \infty.$$

So, there exists a sequence $u_k \in \mathcal{B}_{\varepsilon_k}(U)$ such that $u_k \to y$ in $L^p(\Omega; \mathbb{R}^d)$ and

$$\lim_{k\to\infty} F_{\varepsilon_k}(u_k, U) = F(y, U).$$

Let $\delta > 0$, and then choose $U'$ such that $\operatorname{supp}\psi \subset U' \subset\subset U$ and $|U\backslash U'| \le \delta$. We now use a cut-off argument similarly as in the proof of Lemma 1.23. Fix $N \in \mathbb{N}$ and define

$$U_j = \left\{ x \in U : \operatorname{dist}(x, U') < \frac{j \operatorname{dist}(U', U^c)}{N} \right\}.$$

Then choose the cut-off functions $\varphi_j \in C^\infty_c(U_{j+1}; [0,1])$ with $\varphi_j \equiv 1$ on $U_j$ and $\|\nabla\varphi_j\|_\infty \le CN$ and set

$$\hat{g}_k(x) = g(a), \text{ if } a \in Q_{\varepsilon_k}(x) \cap \mathcal{L}_{\varepsilon_k} \text{ and}$$

$$w_{n,j}(x) = \begin{cases} \varphi_j(\bar{x})u_k(x) + (1 - \varphi_j(\bar{x}))\hat{g}_k(x), & \text{if } x \in U^{\varepsilon_k}, \\ u_k(x) & \text{otherwise.} \end{cases}$$

As in the proof of Lemma 1.23 we calculate

$$\begin{aligned} F_{\varepsilon_k}(w_{k,j}, U) \leq & F_{\varepsilon_k}(u_k, U) + C(\|\nabla g\|_\infty^p + 1)|U \backslash U'| \\ & + \varepsilon_k^d \underbrace{\sum_{\bar{x} \in (\mathcal{L}_{\varepsilon_k}(W_j))^\circ} W_{\text{cell}}(\bar{\nabla} w_{k,j}(\bar{x}))}_{:=S_{j,k}}, \end{aligned}$$

with $W_j = U_{j+2} \backslash \overline{U_{j-1}}$, estimate $S_{j,k}$ by averaging, choose $j(k)$ suitably and thus get

$$\limsup_{k\to\infty} F_{\varepsilon_k}(w_{k,j(k)}, U) \leq F(y, U) + C\delta + \frac{C}{N-4}.$$

Since we choose $j(k) \leq N-3$, we have $w_{k,j(k)} \in \mathcal{B}_{\varepsilon_k}(U, g)$ for any $k$ large enough. Furthermore, $w_{k,j(k)} \to y$ in $L^p(\Omega; \mathbb{R}^d)$ since $\psi$ has support in $U'$. Hence,

$$\Gamma\text{-}\limsup_{k\to\infty} F_{\varepsilon_k}^g(y, U) \leq F^g(y, U) + \delta C + \frac{C}{N-4}.$$

Let $\delta \to 0$ and $N \to \infty$.

In the general case $y|_U \in g + W_0^{1,p}(U; \mathbb{R}^d)$, take $y_l$ such that $y_l|_U \in g + C_c^\infty(U; \mathbb{R}^d)$ and $y_l \to y$ in $W^{1,p}(U; \mathbb{R}^d)$ and in $L^p(\Omega; \mathbb{R}^d)$. We get

$$\begin{aligned} \Gamma\text{-}\limsup_{k\to\infty} F_{\varepsilon_k}^g(y, U) &\leq \liminf_{l\to\infty}(\Gamma\text{-}\limsup_{k\to\infty} F_{\varepsilon_k}^g(y_l, U)) \\ &\leq \liminf_{l\to\infty} F^g(y_l, U) \\ &= F^g(y, U) \end{aligned}$$

by the lower semicontinuity of the Γ-lim sup with respect to $L^p(\Omega; \mathbb{R}^d)$-convergence and continuity of $F^g(\cdot, U)$ with respect to $W^{1,p}(U; \mathbb{R}^d)$-convergence. □

## The Limiting Minimum Problem

The following theorem is important in two ways. On the one hand we gain insight into the Γ-convergence result, on the other hand we will directly need it to get the homogenization result in Section 1.5.

**Theorem 1.28.** *Under the assumptions of Theorem 1.26, we have*

$$\min_y F^g(y, U) = \lim_{k \to \infty} (\inf_y F^g_{\varepsilon_k}(y, U)).$$

*Furthermore, any sequence $y_k$ with equibounded energy is pre-compact in $L^p(U; \mathbb{R}^d)$ and if we have a sequence satisfying*

$$\lim_{k \to \infty} (\inf_y F^g_{\varepsilon_k}(y, U)) = \lim_{k \to \infty} F^g_{\varepsilon_k}(y_k, U),$$

*then every limit of a converging subsequence is a minimizer of $F^g(\cdot, U)$.*

*Proof.* Fix $g, U$ and write $G_k(y) = F^g_{\varepsilon_k}(y, U)$, $G(y) = F^g(y, U)$. Let $y_k$ be a sequence with equibounded energy $G_k(y_k)$. By Assumption 1 and Lemma 1.19 we obtain that

$$\int_{U_{\varepsilon_k}} |\nabla \tilde{y}_k|^p \, dx \le C.$$

Furthermore, using the boundary condition, we have

$$\int_U |\nabla \tilde{y}_k|^p \, dx \le C.$$

A Poincaré-type inequality involving the trace yields

$$\|\tilde{y}_k\|_{W^{1,p}(U;\mathbb{R}^d)} \le C(\|\nabla \tilde{y}_k\|_{L^p(U;\mathbb{R}^d)} + \|T\tilde{y}_k\|_{L^p(\partial U;\mathbb{R}^d)})$$
$$\le C + C\|g\|_\infty \mathcal{H}^{d-1}(\partial U)^{\frac{1}{p}} \le C$$

and so $\tilde{y}_{k_l} \to y$ in $L^p(U; \mathbb{R}^d)$ for some subsequence $k_l$ and some $y \in W^{1,p}(U; \mathbb{R}^d)$. Then, by Lemma 1.27, $y_{k_l} \to y$ in $L^p(U; \mathbb{R}^d)$.

Now from Theorem 1.26 we infer that $G_k$ $\Gamma(L^p(\Omega; \mathbb{R}^d))$-converges to $G$. But then $G_k$ also $\Gamma(L^p(U; \mathbb{R}^d))$-converges to $G$. Here the existence of recovery sequences is immediate as $L^p(\Omega; \mathbb{R}^d)$- implies $L^p(U; \mathbb{R}^d)$-convergence. As for the liminf-inequality, if $y_k \to y$ in $L^p(U; \mathbb{R}^d)$, where the energies $G_k(y_k)$ are without loss of generality assumed to be equibounded and, in particular, in $y_k \in \mathcal{B}_{\varepsilon_k}(U, g)$, we can extend the functions by defining $y_k(x) := g(\bar{x})$ and $y(x) := g(x)$ for $x \in \Omega \backslash U$ without changing their respective energies. Since then $y_k \to y$ in $L^p(\Omega; \mathbb{R}^d)$, we have indeed that $\liminf_{k \to \infty} G_k(y_k) \ge G(y)$. The remaining part of the proof now directly follows from Theorem 1.14. $\square$

## 1.5    Proof of the Main Results

### The Γ-Convergence Results

To simplify notations, we define $P_h(x) = x + A(0,h)^d$ and $P_h = P_h(0)$. First, we will prove the following lemma.

**Lemma 1.29.** *The limit*

$$\lim_{N\to\infty} \frac{1}{N^d} \inf\left\{ \sum_{x\in(\mathcal{L}_1'(P_N))^\circ} W_{\text{cell}}(\bar{\nabla}y(x))\colon y \in \mathcal{B}_1(P_N, y_M) \right\}$$

*exists for every $M \in \mathbb{R}^{d\times d}$.*

*Proof.* Let us define

$$G(y, U) = \sum_{x\in(\mathcal{L}_1'(U))^\circ} W_{\text{cell}}(\bar{\nabla}y(x)) \quad \text{and}$$

$$f_k(M) = \frac{1}{k^d} \inf\left\{ G(y, P_k)\colon y \in \mathcal{B}_1(P_k, y_M) \right\}.$$

Fix $M \in \mathbb{R}^{d\times d}$ and let $k, n \in \mathbb{N}$ with $k > n$. Choose $v_n \in \mathcal{B}_1(P_n, y_M)$ such that

$$\frac{1}{n^d} G(v_n, P_n) \le f_n(M) + \frac{1}{n}.$$

Allowing $\alpha \in A\left\{0, \dots, \left[\frac{k}{n}\right] - 1\right\}^d$ we define

$$u_k(x) = \begin{cases} v_n(x - n\alpha) + nM\alpha & \text{if } x \in P_n(n\alpha) \text{ for some } \alpha, \\ M\bar{x} & \text{otherwise.} \end{cases}$$

Since $v_n$ satisfies the boundary condition, $u_k$ is constant on every cell.

Moreover, $u_k \in \mathcal{B}_1(P_k, y_M)$ and we can estimate

$$
\begin{aligned}
f_k(M) &\leq \frac{1}{k^d} G(u_k, P_k) \\
&\leq \frac{1}{k^d} \left( c(|M|^p + 1) \left( \#(\mathcal{L}_1'(P_k))^\circ - \left[\frac{k}{n}\right]^d \#(\mathcal{L}_1'(P_n(\alpha n)))^\circ \right) \right. \\
&\quad \left. + \left[\frac{k}{n}\right]^d G(v_n, P_n) \right) \\
&\leq \frac{c(|M|^p + 1)}{k^d} \left( \frac{|P_k| - |P_{n\left[\frac{k}{n}\right]}|}{|P_1|} + \left[\frac{k}{n}\right]^d \left( n^d - (n-2)^d \right) \right) \\
&\quad + \frac{1}{n^d} G(v_n, P_n) \\
&\leq f_n(M) + \frac{1}{n} + \frac{c(|M|^p + 1)}{k^d} \left( k^d - \left( n\left[\frac{k}{n}\right] \right)^d + k^d \left( 1 - \left( 1 - \frac{2}{n} \right)^d \right) \right) \\
&\leq f_n(M) + \frac{1}{n} + c(|M|^p + 1) \left( 1 - \left( 1 - \frac{n}{k} \right)^d + 1 - \left( 1 - \frac{2}{n} \right)^d \right).
\end{aligned}
$$

Thus, for every $n \in \mathbb{N}$,

$$
\limsup_{k \to \infty} f_k(M) \leq f_n(M) + \frac{1}{n} + c(|M|^p + 1) \left( 1 - \left( 1 - \frac{2}{n} \right)^d \right),
$$

hence,

$$
\limsup_{k \to \infty} f_k(M) \leq \liminf_{n \to \infty} f_n(M).
$$

$\square$

Now, we can prove our first main theorem.

*Proof of Theorem 1.1.* We will first show that $F_\varepsilon(\cdot, \Omega)$ $\Gamma(L^p(\Omega; \mathbb{R}^d))$-converges to $F$. According to Lemma 1.29, $W_{\text{cont}}$ is well-defined. By the Urysohn property of $\Gamma$-convergence in Proposition 1.13, it is enough to show that, for any sequence $\varepsilon_n \to 0$, the function $f$ of Theorem 1.18 equals $W_{\text{cont}}$. Fix such a sequence, the subsequence $\varepsilon_k$ and the associated $f$. Since $f$ is quasiconvex, we have for every $M \in \mathbb{R}^{d \times d}$ and $U \in \mathcal{A}_L(\Omega)$

$$
\begin{aligned}
f(M) &= \frac{1}{|U|} \min \left\{ \int_U f(\nabla y(x)) \, dx : y - y_M \in W_0^{1,p}(U; \mathbb{R}^d) \right\} \\
&= \frac{1}{|U|} \min \left\{ F(y, U) : y - y_M \in W_0^{1,p}(U; \mathbb{R}^d) \right\}.
\end{aligned}
$$

If we restrict $y_M$ to a ball that contains some neighborhood of $\Omega$, we can extend it to a function in $W^{1,\infty}(\mathbb{R}^d; \mathbb{R}^d) \cap C(\mathbb{R}^d; \mathbb{R}^d)$, so $y_M$ is admissible as a boundary condition in Theorem 1.26 and we get the Γ-convergence result with boundary condition. Hence by Theorem 1.28, for $h_0 > 0$ and $x_0 \in \mathbb{R}^d$ such that $P_{h_0}(x_0) \subset\subset \Omega$,

$$
f(M) = \frac{1}{|P_{h_0}(x_0)|} \lim_{k\to\infty} \left( \inf \left\{ F_{\varepsilon_k}(y, P_{h_0}(x_0)) \colon y \in \mathcal{B}_{\varepsilon_k}(P_{h_0}(x_0), y_M) \right\} \right)
$$

$$
= \frac{1}{|P_{h_0}(x_0)|} \lim_{k\to\infty} \inf \left\{ \varepsilon_k^d \sum_{x \in (\mathcal{L}'_{\varepsilon_k}(P_{h_0}(x_0)))^\circ} W_{\text{cell}}(\bar{\nabla} y(x)) \colon \right.
$$

$$
\left. y \in \mathcal{B}_{\varepsilon_k}(P_{h_0}(x_0), y_M) \right\}.
$$

It is easy to see, that we can always find $h_k > 0$ and $x_k \in \mathcal{L}_{\varepsilon_k}$ such that

$$
P_{h_k}(x_k) = \left( \bigcup_{x \in P_{h_0}(x_0)} Q_{\varepsilon_k}(x) \right)^\circ .
$$

We then know $P_{h_k}(x_k) \subset \Omega$ for all $k$ large enough,

$$
|x_0 - x_k| \leq \operatorname{diam} Q_{\varepsilon_k} = \varepsilon_k \operatorname{diam} Q_1,
$$

$h_0 \leq h_k \leq h_0 + 2\varepsilon_k$, and, that there are $N_k \in \mathbb{N}$ satisfying $h_k = N_k \varepsilon_k$. Furthermore,

$$
\mathcal{L}'_{\varepsilon_k}(P_{h_0}(x_0)) = \mathcal{L}'_{\varepsilon_k}(P_{h_k}(x_k)) \quad \text{and}
$$

$$
(\mathcal{L}'_{\varepsilon_k}(P_{h_0}(x_0)))^\circ = (\mathcal{L}'_{\varepsilon_k}(P_{h_k}(x_k)))^\circ.
$$

Hence, $\mathcal{B}_{\varepsilon_k}(P_{h_0}(x_0), y_M)$ and $\mathcal{B}_{\varepsilon_k}(P_{h_k}(x_k), y_M)$ are equal up to extending the functions in $\mathcal{B}_{\varepsilon_k}(P_{h_0}(x_0), y_M)$ constant on cells that intersect $P_{h_k}(x_k) \setminus P_{h_0}(x_0)$. It follows that

$$
f(M) = \frac{1}{|P_1|} \lim_{k\to\infty} \frac{1}{N_k^d} \frac{h_k^d}{h_0^d} \inf \left\{ \sum_{x \in (\mathcal{L}'_{\varepsilon_k}(P_{h_k}(x_k)))^\circ} W_{\text{cell}}(\bar{\nabla} y(x)) \colon \right.
$$

$$
\left. y \in \mathcal{B}_{\varepsilon_k}(P_{h_k}(x_k), y_M) \right\}
$$

$$
= \frac{1}{|P_1|} \lim_{k\to\infty} \frac{1}{N_k^d} \inf \left\{ \sum_{x \in (\mathcal{L}'_{\varepsilon_k}(P_{h_k}))^\circ} W_{\text{cell}}(\bar{\nabla} y(x)) \colon y \in \mathcal{B}_{\varepsilon_k}(P_{h_k}, y_M) \right\},
$$

where we used, that $y \in \mathcal{B}_{\varepsilon_k}(P_{h_k}(x_k), y_M)$, if and only if $y(\cdot + x_k) - Mx_k \in \mathcal{B}_{\varepsilon_k}(P_{h_k}, y_M)$ and that the discrete gradient of $y$ at a point $x$ equals the discrete gradient of $y(\cdot + x_k) - Mx_k$ at $x - x_k$. In a similar way $y \in \mathcal{B}_{\varepsilon_k}(P_{h_k}, y_M)$ if and only if $y' \in \mathcal{B}_1(P_{N_k}, y_M)$ and $\bar{\nabla}y'(x) = \bar{\nabla}y(\varepsilon_k x)$, where $y'(x) = \frac{1}{\varepsilon_k}y(\varepsilon_k x)$. Hence,

$$f(M) = \frac{1}{|P_1|} \lim_{k \to \infty} \frac{1}{N_k^d} \inf \left\{ \sum_{x \in (\mathcal{L}_1'(P_{N_k}))^\circ} W_{\text{cell}}(\bar{\nabla}y(x)) \colon y \in \mathcal{B}_1(P_{N_k}, y_M) \right\}$$
$$= W_{\text{cont}}(M).$$

In order to prove that also $F_\varepsilon(\cdot, \Omega)$ $\Gamma(L_{loc}^p(\Omega; \mathbb{R}^d)/\mathbb{R})$-converges to $F$, we only need to verify the lim inf-inequality as the existence of recovery sequences immediately follows from the first part of the proof since convergence in $L^p(\Omega; \mathbb{R}^d)$ implies convergence in $L_{loc}^p(\Omega; \mathbb{R}^d)/\mathbb{R}$. But if $\varepsilon_n \to 0$ and $y_{\varepsilon_n} \to y$ in $L_{loc}^p(\Omega; \mathbb{R}^d)/\mathbb{R}$, then there exist $c_n \in \mathbb{R}$ such that, for every $U \in \mathcal{A}_L(\Omega)$ with $U \subset\subset \Omega$, $y_{\varepsilon_n} - c_n \to y$ in $L^p(U; \mathbb{R}^d)$, so that by the previous result

$$\liminf_{n \to \infty} F_{\varepsilon_n}(y_{\varepsilon_n}, \Omega) = \liminf_{n \to \infty} F_{\varepsilon_n}(y_{\varepsilon_n} - c_n, \Omega)$$
$$\geq \liminf_{n \to \infty} F_{\varepsilon_n}(y_{\varepsilon_n} - c_n, U)$$
$$\geq F(y, U).$$

Without loss of generality we may assume that $\liminf_{k \to \infty} F_{\varepsilon_k}(y_{\varepsilon_k}, \Omega) < \infty$. Since for any $V \in \mathcal{A}(\Omega)$ with $V \subset\subset \Omega$ there exists $U \in \mathcal{A}_L(\Omega)$ with $V \subset U \subset\subset \Omega$, we then deduce from Lemma 1.21 that $y \in W^{1,p}(V; \mathbb{R}^d)$ with $\|y\|_{W^{1,p}(V;\mathbb{R}^d)}$ bounded uniformly in $V \in \mathcal{A}$ with $V \subset\subset \Omega$, hence $y \in W^{1,p}(\Omega; \mathbb{R}^d)$. Then invoking Lemma 1.24 and passing to the supremum over $U \in \mathcal{A}_L(\Omega)$ in the above inequality yields

$$\liminf_{k \to \infty} F_{\varepsilon_k}(y_{\varepsilon_k}, \Omega) \geq F(y, \Omega).$$

$\square$

*Proof of Theorem 1.4.* Theorem 1.4 is a direct consequence of Theorem 1.26 and Theorem 1.1, where the limiting energy density $f$ has been identified as $W_{\text{cont}}$. $\square$

*Proof of Theorem 1.3.* Suppose $y_k$ is a sequence with equibounded energies $F_{\varepsilon_k}(y_k)$. By Lemma 1.19 and the growth assumptions on $W_{\text{cell}}$, for every $U \in \mathcal{A}(\Omega)$ with $U \subset\subset \Omega$ we have

$$\int_U |\nabla \tilde{y}_k|^p \leq C F_{\varepsilon_k}(y_k) + C|\Omega|$$

uniformly bounded for sufficiently large $k$. Choose $U_0 \in \mathcal{A}_L(\Omega)$ connected and with $\emptyset \neq U_0 \subset\subset \Omega$. As $U_0$ is connected, by Poincaré's inequality we find $c_k \in \mathbb{R}$ such that $\tilde{y}_k - c_k$ is pre-compact in $L^p(U_0; \mathbb{R}^d)$. But then indeed for any connected $U \in \mathcal{A}_L$ with $U_0 \subset U \subset\subset \Omega$ the Poincaré inequality

$$\|\tilde{y}_k - c_k\|_{W^{1,p}(U; \mathbb{R}^d)} \leq C \|\nabla \tilde{y}_k\|_{L^p(U; \mathbb{R}^d)} + \|\tilde{y}_k - c_k\|_{L^p(U_0; \mathbb{R}^d)}$$

yields that $\tilde{y}_k - c_k$ is pre-compact in $L^p(U; \mathbb{R}^d)$. Exhausting $\Omega$ with a countable number of such domains and passing to a diagonal sequence, we find a subsequence $y_{k_n}$ such that $\tilde{y}_{k_n} - c_{k_n}$ converges in $L^p_{loc}(\Omega; \mathbb{R}^d)$. By Lemma 1.20 we finally obtain that $y_{k_n} - c_{k_n}$ converges in $L^p_{loc}(\Omega; \mathbb{R}^d)$.  □

*Proof of Theorem 1.5.* This is immediate from Theorem 1.28.  □

*Proof of Corollary 1.6.* This is a direct consequence of Theorems 1.1, 1.3, 1.4, 1.5 and 1.14.  □

*Proof of Theorem 1.9.* If in addition to Assumptions 1 and 2 Assumption 3 holds true, we can apply [CDKM06, Theorem 4.2] with $\Lambda = (\mathcal{L}'_1(P_{N_k}))^\circ$. It is easy to see that the boundary of $\Lambda$ as defined in [CDKM06] equals $\partial \mathcal{L}_1(P_{N_k}) \cup \mathcal{L}_1 \backslash \mathcal{L}'_1(P_{N_k})$, but of course the second part does not change anything. This shows that there is a neighborhood $\mathcal{U}$ of $SO(d)$, such that for every $M \in \mathcal{U}$

$$W_{\text{cont}}(M) = \frac{1}{|P_1|} \lim_{k \to \infty} \frac{1}{N_k^d} \inf \left\{ \sum_{x \in (\mathcal{L}'_1(P_{N_k}))^\circ} W_{\text{cell}}(\bar{\nabla} y(x)) : \right.$$

$$\left. y \in \mathcal{B}_1(P_{N_k}, y_M) \right\}$$

$$= \frac{1}{|\det A|} \lim_{k \to \infty} \frac{1}{N_k^d} \sum_{x \in (\mathcal{L}'_1(P_{N_k}))^\circ} W_{\text{cell}}(MZ)$$

$$= \frac{1}{|\det A|} W_{\text{cell}}(MZ)$$

$$= W_{\text{CB}}(M).$$

□

## Approximation of General Continuum Densities

Next we prove Propositions 1.10 and 1.11.

*Proof of Proposition 1.10.* At variance with our previous decomposition procedure, we now choose any simplicial decomposition $\mathcal{S}$ of the cell $A[-\frac{1}{2}, \frac{1}{2})^d$ into $d$-simplices all of whose corners lie in $A\{-\frac{1}{2}, \frac{1}{2}\}^d$. For $F = (f_1, \ldots, f_{2^d}) \in \mathbb{R}^{d \times 2^d}$ we then interpolate the mapping

$$A\left\{-\frac{1}{2}, \frac{1}{2}\right\}^d \to \mathbb{R}^d, \qquad x_i \mapsto f_i$$

affine on each simplex in order to obtain

$$u_F : A\left[-\frac{1}{2}, \frac{1}{2}\right)^d \to \mathbb{R}^d.$$

Then $W_{\text{cell}}$ is defined by

$$W_{\text{cell}}(F) := \int_{A[-\frac{1}{2}, \frac{1}{2})^d} V(\nabla u_F)\, dx.$$

As every corner $z_{i_0}, \ldots, z_{i_d}$ of $S \in \mathcal{S}$ lies in $A\{-\frac{1}{2}, \frac{1}{2}\}^d$, we have

$$c \sum_{j=1}^{d} |f_{i_j} - f_{i_0}| \leq |\nabla u_F| \leq C|F|$$

on $S$. Thus, $\|F\| = \max\limits_{x \in A[-\frac{1}{2}, \frac{1}{2})^d} |\nabla u_F(x)|$ is a norm on $V_0$ and we calculate

$$
\begin{aligned}
W_{\text{cell}}(F) &\geq \int_{A[-\frac{1}{2}, \frac{1}{2})^d} c|\nabla u_F|^p - c'\, dx \\
&\geq c\|F\|^p - c' \\
&\geq c|F|^p - c',
\end{aligned}
$$

and on the other hand

$$W_{\text{cell}}(F) \leq C(|F|^p + 1).$$

This means $W_{\text{cell}}$ satisfies Assumptions 1 and 2. From Theorem 1.1 we then deduce that

$$
W_{\text{cont}}(M) = \frac{1}{|\det A|} \lim_{N \to \infty} \frac{1}{N^d} \inf \left\{ \sum_{x \in (\mathcal{L}'_1(P_N))^{\circ}} W_{\text{cell}}(\bar{\nabla} y(x)) : \right.
$$
$$
\left. y \in \mathcal{B}_1(P_N, y_M) \right\}
$$
$$
= \frac{1}{|\det A|} \lim_{N \to \infty} \frac{1}{N^d} \inf \left\{ \int_{(P_N)_1} V(\nabla u_{\bar{\nabla} y(x)}) : y \in \mathcal{B}_1(P_N, y_M) \right\}
$$
$$
= \frac{1}{|\det A|} \lim_{N \to \infty} \frac{1}{N^d} |\det A| (N-2)^d V(M)
$$
$$
= V(M)
$$

due to the quasiconvexity of $V$.                                         □

*Proof of Proposition 1.11.* Any $F \in V_0$ can be decomposed orthogonally as $F = F'Z + F''$ with unique $F' \in \mathbb{R}^{d \times d}$ and $F'' \in (\mathbb{R}^{d \times d} Z)^{\perp}$. Set

$$
W_{\text{cell}}(F) = |\det A| Q \left( \sqrt{(F')^T F'} - \text{Id} \right) + |F''|^2 + \chi(F),
$$

where $\chi$ is any frame indifferent function satisfying Assumptions 1 and 2 with $p \geq d$ which is non-negative, vanishes near $\bar{SO}(d)$ and is bounded from below by a positive constant on $\bar{O}(d) \setminus \bar{SO}(d)$, $\bar{O}(d) = O(d)Z$. Then also $W_{\text{cell}}$ satisfies Assumptions 1 and 2 with the same $p$. Noting that, for $M \in \mathbb{R}^{d \times d}$, $(MF)' = MF'$ and $(MF)'' = MF''$, it is not hard to verify that $W_{\text{cell}}$ also satisfies Assumption 3 with

$$
D^2 W_{\text{cell}}(Z)(F, F) = 2|\det A| Q \left( \frac{(F')^T + F'}{2} \right) + 2|F''|^2.
$$

But then

$$
\frac{1}{2} D^2 W_{\text{CB}}(\text{Id})(M, M) = \frac{1}{2|\det A|} D^2 W_{\text{cell}}(Z)(MZ, MZ)
$$
$$
= Q \left( \frac{M^T + M}{2} \right)
$$
$$
= Q(M).
$$

                                                                          □

## Extension to Finite-Range Energies

We briefly comment on more general long-range interactions. Suppose we have a fixed finite set $\Lambda = \{z_1, \ldots, z_{2^d}, \ldots, z_N\} \subset \mathcal{L}'$, where $z_1, \ldots, z_{2^d}$ still denote $A\{-\frac{1}{2}, \frac{1}{2}\}^d$. For $y \in \mathcal{B}_\varepsilon(\Omega)$ we define $y_i = y(\bar{x} + \varepsilon z_i)$. With $\bar{x}$ and $\bar{y}$ as before, i.e., only depending on $y_1, \ldots y_{2^d}$, let now

$$\bar{\nabla} y(x) = \frac{1}{\varepsilon}(y_1 - \bar{y}, \ldots, y_N - \bar{y}) \in \mathbb{R}^{d \times N}.$$

The lattice interior $(\mathcal{L}'_\varepsilon(U))^\circ$ and boundary $\partial \mathcal{L}'_\varepsilon(U)$ now have to be shrunk respectively enlarged to a whole boundary layer, according to the maximal interaction length in $\Lambda$. Assumptions 1 and 2 are then replaced by the estimate

$$c|F'|^p - c' \leq W_{\text{super-cell}}(F) \leq c''(|F|^p + 1)$$

for constants $c, c', c'' > 0$ and all $F \in \mathbb{R}^{d \times N}$ which satisfy $F' \in V_0$, where $F' \in \mathbb{R}^{d \times 2^d}$ denotes the left $d \times 2^d$ submatrix of $F$. Note that the lower bound in particular allows for arbitrarily weak long range interactions. As the interpolation we used only depends on the $d \times 2^d$ values of the corresponding lattice cell, this implies that we get the standard estimates for the gradients in Lemma 1.19 only on this part of the discrete gradient.

It is important that the interaction range is bounded by $C\varepsilon$, so that, e.g., Lemma 1.22 and its proof still work. In the estimates of the error $S_{j,n}$ in, e.g., Lemma 1.23, it is important that $\varepsilon^{-1}|u(\bar{x} + \varepsilon z_i) - u(\bar{x})|$ and thus the discrete gradient can be bounded by a fixed finite sum of smaller $d \times 2^d$ discrete gradients of some cells near $x$. Hence, we still have the estimate

$$\varepsilon_n^d \sum_{x \in (\mathcal{L}'_{\varepsilon_n}(U))^\circ} |\bar{\nabla} u(x)|^p \leq C \int_U |\nabla \tilde{u}(x)|^p \, dx$$

Note that according to our enlarging of the lattice boundaries, also the cell formula for the limit density will now involve a sequence of minimizing problems with affine boundary conditions on a boundary layer.

We finally remark that the statement on the applicability of the Cauchy-Born rule translates naturally, as the main ingredient does, see [CDKM06, Theorem 5.1].

## Extension to Multi-Lattices

It is also possible to generalize these results to certain non-Bravais lattices, namely to multi-lattices of the form $\mathcal{L} \cup (s_1 + \mathcal{L}) \cup \cdots \cup (s_m + \mathcal{L})$, in the following way: We still consider $\mathcal{L}$ to be our main lattice. But now we have $m$ additional atoms in each cell, which we describe by the 'internal variable' $s(x) \in \mathbb{R}^{d \times m}$, such that $\varepsilon s_{.j}$ describes the distance of the $j$-th atom to the midpoint of the cell. Of course $s$ can be identified with a function, that is constant on every interior cell and is 0 outside and thus lies in some $L^q(\Omega; \mathbb{R}^{d \times m})$, $1 < q < \infty$. The new cell energy depends on $md$ additional variables and we now consider the growth condition

$$c(|M'|^p + |s|^q) - c' \leq W_{\text{super-cell}}(M, s) \leq c''(|M|^p + |s|^q + 1)$$

for $M \in \mathbb{R}^{d \times N}$ and $s \in \mathbb{R}^{d \times m}$. It is now natural to have a Γ-convergence result with respect to strong-$L^p$-convergence in the first and weak-$L^q$-convergence in the second component. As we will see in a moment, it turns out that we have to consider a combined boundary value and mean value problem. For this we define $\mathcal{B}_\varepsilon(U, g, s_0)$ to consist of all pairs $(y, s)$, such that $y \in \mathcal{B}_\varepsilon(U, g)$ and $s \in L^q(\Omega; \mathbb{R}^{d \times m})$ is constant on every interior cell of $U$, is 0 outside and has mean value $s_0$ on the union of interior cells of $U$.

In analogy to Theorem 1.1, we now have Theorem 1.7. The proof of this theorem is similar to the proof of Theorem 1.1. But there are several things that need to be addressed:

First of all, the weak topology on $L^q$ is not given by a metric. But, as discussed in [DM93] in detail, this is not a big problem, since our functionals are equicoercive and the dual of $L^q$ is separable. In particular, we can describe Γ-convergence by sequences and the compactness and the Urysohn property are still true. Next, we need an advanced version of our integral representation result in the so called cross-quasiconvex case.

**Theorem 1.30.** *Let $1 \leq p, q < \infty$ and let*

$$F \colon W^{1,p}(\Omega; \mathbb{R}^d) \times L^q(\Omega; \mathbb{R}^{d \times m}) \times \mathcal{A}(\Omega) \to [0, \infty]$$

*satisfy the following conditions:*

(i) *(locality) $F(y, s, U) = F(v, t, U)$, if $y(x) = v(x)$ and $s(x) = t(x)$ for a.e. $x \in U$;*

(ii) *(measure property)* $F(y, s, \cdot)$ *is the restriction of a Borel measure to* $\mathcal{A}(\Omega)$;

(iii) *(growth condition) there exists* $c > 0$ *such that*

$$F(y, s, U) \leq c \int_U |\nabla y(x)|^p + |s|^q + 1 \, dx;$$

(iv) *(translation invariance in y)* $F(y, s, U) = F(y + a, s, U)$ *for every* $a \in \mathbb{R}^d$ ;

(v) *(lower semicontinuity)* $F(\cdot, \cdot, U)$ *is sequentially lower semicontinuous with respect to weak convergence in* $W^{1,p}(\Omega; \mathbb{R}^d)$ *in the first and weak convergence in* $L^q(\Omega; \mathbb{R}^{d \times m})$ *in the second component;*

(vi) *(translation invariance in x)* With $y_M(x) = Mx$ *and* $s(x) = s_0$ *we have*

$$F(y_M, s, B_r(x)) = F(y_M, s, B_r(x'))$$

*for every* $M \in \mathbb{R}^{d \times d}$, $s_0 \in \mathbb{R}^{d \times m}$, $x, x' \in \Omega$ *and* $r > 0$ *such that* $B_r(x) \subset \Omega$ *and* $B_r(x') \subset \Omega$.

*Then there exists a continuous* $f \colon \mathbb{R}^{d \times d} \times \mathbb{R}^{d \times m} \to [0, \infty)$ *such that*

$$0 \leq f(M, s) \leq C(1 + |M|^p + |s|^q)$$

*for every* $M \in \mathbb{R}^{d \times d}$, $s \in \mathbb{R}^{d \times m}$ *and*

$$F(y, s, U) = \int_U f(\nabla y(x), s(x)) \, dx$$

*for every* $y \in W^{1,p}(\Omega; \mathbb{R}^d)$, $s \in L^q(\Omega; \mathbb{R}^{d \times m})$ *and* $U \in \mathcal{A}(\Omega)$.

The proof in [BD98] for the pure Sobolev version of this theorem readily applies to this more general statement. (Note that continuity of $f$ then follows from separate convexity.) Most of the lemmata then translate naturally. We just want to comment on some details in Lemma 1.23. The recovery sequences now contain additionally some $t_n \rightharpoonup s$, $r_n \rightharpoonup s$ corresponding to $U, V$ respectively. We define

$$q_{n,j} = \chi_{U_j}(\bar{x}) t_n(x) + (1 - \chi_{U_j}(\bar{x})) r_n(x),$$

and then choose $j(n)$ as before to define $s_n = q_{n,j(n)}$. The only part that is not immediately clear now, is the convergence $s_n \rightharpoonup s$. To prove this, let $\varphi \in L^{q'}(\Omega; \mathbb{R}^{d \times m})$. We now split $\varphi$ into several parts we can control

$$\varphi = \psi_n + \varphi \chi_{U'} + \varphi \chi_{\Omega \setminus \overline{U_N}} + \sum_{j=0}^{N-1} \varphi \chi_{(U_{j+1} \setminus \overline{U_j})_{\varepsilon_n}}.$$

Here the $\psi_n$ contain all the remaining parts. We see that $\psi_n \to 0$ strongly in $L^{q'}$ as long as $|\partial U_j| = 0$ for every $j$, so this is true up to changing the sets $U_j$ a little bit. But then we also have

$$\varphi \chi_{(U_{j+1} \setminus \overline{U_j})_{\varepsilon_n}} \to \varphi \chi_{U_{j+1} \setminus U_j}$$

strongly in $L^{q'}$. The advantage is now that on each set $(U_{j+1} \setminus \overline{U_j})_{\varepsilon_n}$ we have either $s_n = t_n$ or $s_n = r_n$, possibly changing with $n$. But in both cases we have weak convergence to $s$, hence

$$\int_\Omega s_n(x)\varphi(x)\chi_{(U_{j+1} \setminus \overline{U_j})_{\varepsilon_n}}(x)\,dx \to \int_\Omega s(x)\varphi(x)\chi_{U_{j+1} \setminus U_j}(x)\,dx.$$

And, putting it all together, we get

$$\int_\Omega s_n(x)\varphi(x)\,dx \to \int_\Omega s(x)\varphi(x)\,dx.$$

Another important step is to adjust Theorem 1.26 and Theorem 1.28 so that additionally to the boundary values for $y$ we have a fixed mean value for $s$, i.e., we consider $\mathcal{B}_\varepsilon(U, g, s_0)$ instead of $\mathcal{B}_\varepsilon(U, g)$ in the discrete setting and add the constraint

$$\fint_U s(x)\,dx = s_0$$

in the continuum setting. To get the lim inf-inequality just notice that for $s_k \rightharpoonup s$ with $(y_k, s_k) \in \mathcal{B}_\varepsilon(U, g, s_0)$ we have

$$\fint_U s(x)\,dx = \lim_{k\to\infty} \fint_U s_k(x)\,dx$$
$$= \lim_{k\to\infty} \frac{|U_{\varepsilon_k}|}{|U|} \fint_{U_{\varepsilon_k}} s_k(x)\,dx$$
$$= \lim_{k\to\infty} \frac{|U_{\varepsilon_k}|}{|U|} s_0 = s_0.$$

The lim sup-inequality is a little more subtle. We have a function $s \in L^q(\Omega; \mathbb{R}^{d \times m})$ with

$$\fint_U s(x)\, dx = s_0$$

and a recovery sequence without this mean value $s_k \rightharpoonup s$. Let us write

$$\fint_{U_{\varepsilon_k}} s_k(x)\, dx + \xi_k = s_0,$$

so that $\xi_k \to 0$. We now adjust the $s_k$ adequately. Define

$$t_k(x) = s_k(x) + \xi_k \frac{|U_{\varepsilon_k}|}{|V_k|} \chi_{V_k}.$$

If $V_k$ is a union of cells with some distance to the boundary of $U$, then, for $k$ large enough, the $t_k$ are admissible functions and do not interact with the adjustments on $y$. We have to make sure, that $t_k \rightharpoonup s$ and

$$\limsup_{k \to \infty} F_{\varepsilon_k}(u_k, t_k, U) \leq \limsup_{k \to \infty} F_{\varepsilon_k}(u_k, s_k, U).$$

The weak convergence is true, if $|V_k| \to 0$ and $|V_k| \geq c\xi_k$ for some $c > 0$. For the second estimate, we have to choose the $V_k$ a little more carefully to avoid concentration of the energy. Choose sequences $\eta_k \to 0$, $L_k \to \infty$ such that $\eta_k \geq c\xi_k$, $\frac{\eta_k}{\varepsilon_k^d} \to \infty$ and $L_k \eta_k \to 0$. Then take $L_k \in \mathbb{N}$ disjoints sets $W_{k,l} \subset U$, that are unions of cells, such that $|W_{k,l}|$ is independent of $l$ and is roughly equal to $\eta_k$, which means

$$c\eta_k \leq |W_{k,l}| \leq C\eta_k,$$

with $C, c > 0$ independent of $k$ and $l$. This is possible as $\frac{\eta_k}{\varepsilon_k^d} \to \infty$ and $L_k \eta_k \to 0$. Then, we can choose $l(k)$ and set $V_k = W_{k,l(k)}$, such that

$$\int_{V_k} W_{\mathrm{cell}}(\bar{\nabla} u_k(x), s_k(x)) + W_{\mathrm{cell}}\left(\bar{\nabla} u_k(x), s_k(x) + \xi_k \frac{|U_{\varepsilon_k}|}{|V_k|}\right) dx$$

$$\leq \frac{1}{L_k} \sum_{l=1}^{L_k} \int_{W_{k,l}} W_{\mathrm{cell}}(\bar{\nabla} u_k(x), s_k(x))$$

$$+ W_{\mathrm{cell}}\left(\bar{\nabla} u_k(x), s_k(x) + \xi_k \frac{|U_{\varepsilon_k}|}{|W_{k,l}|}\right) dx$$

$$\leq \frac{1}{L_k} C,$$

due to the growth condition. So the error goes to zero with $L_k \to \infty$. The rest of the proof translates naturally. The most important observation is the equality

$$f(M, s_0) = \frac{1}{|U|} \min \left\{ \int_U f(\nabla y(x), s(x)) \, dx \colon y - y_M \in W_0^{1,p}(U; \mathbb{R}^d), \right.$$

$$\left. s \in L^q(U; \mathbb{R}^{d \times m}), \fint_U s(x) \, dx = s_0 \right\}$$

$$= \frac{1}{|U|} \min \left\{ F(y, s, U) \colon y - y_M \in W_0^{1,p}(U; \mathbb{R}^d), \right.$$

$$\left. s \in L^q(U; \mathbb{R}^{d \times m}), \fint_U s(x) \, dx = s_0 \right\},$$

which is of course a consequence of the lower semicontinuity properties.

*Proof of Theorem 1.8.* Fix $y \in W^{1,p}(\Omega; \mathbb{R}^d)$ and without loss of generality fix a version of $y$ that is finite everywhere.

Due to the growth condition and the continuity, we know that the infimum in

$$\inf_{s \in \mathbb{R}^{d \times m}} W_{\text{cont}}(M, s)$$

is actually a minimum for arbitrary $M$ and that the function

$$M \mapsto \min_{s \in \mathbb{R}^{d \times m}} W_{\text{cont}}(M, s)$$

is continuous. Obviously, we always have the inequality

$$\int_\Omega W_{\text{cont}}(\nabla y(x), s(x)) \, dx \geq \int_\Omega \min_{s \in \mathbb{R}^{d \times m}} W_{\text{cont}}(\nabla y(x), s) \, dx.$$

We now want to show, that there always exists an $L^q$-function $s$ where this is an equality. The idea is of course to choose $s(x)$ as a minimizer of $s \mapsto W_{\text{cont}}(\nabla y(x), s)$. The key point is to ensure measurability. We will do this by using the theory of measurable multifunctions as developed, e.g., in [FL07]. Define

$$\Theta(M) = \{ s \in \mathbb{R}^{d \times m} \colon W_{\text{cont}}(M, s) = \min_{t \in \mathbb{R}^{d \times m}} W_{\text{cont}}(M, t) \}$$

and set $\Gamma(x) = \Theta(\nabla y(x))$. Due to the continuity and the growth of $W_{\text{cont}}$, the set $\Gamma(x)$ is closed and non-empty for every $x \in \Omega$, hence $\Gamma \colon \Omega \to \mathcal{P}(\mathbb{R}^{d \times m})$ is a closed-valued multifunction.

Next, we want to show that $\Gamma$ is measurable, in the sense that

$$\Gamma^-(C) = \{x \in \Omega \colon \Gamma(x) \cap C \neq \emptyset\}$$

is Lebesgue-measurable for every closed set $C \subset \mathbb{R}^{d \times m}$. To this end, we will first show that $\Theta^-(C)$ is closed. Let $M_n \in \Theta^-(C)$, $M_n \to M$ and choose $s_n \in \Theta(M_n) \cap C$. Using the growth of $W_{\text{cont}}$ and since the $M_n$ are bounded, the $s_n$ are also bounded. So for some subsequence we have $s_{n_k} \to s$ and $s \in C$. Furthermore,

$$\begin{aligned} W_{\text{cont}}(M, s) &= \lim_{k \to \infty} W_{\text{cont}}(M_{n_k}, s_{n_k}) \\ &= \lim_{k \to \infty} \min_{t \in \mathbb{R}^{d \times m}} W_{\text{cont}}(M_{n_k}, t) \\ &= \min_{t \in \mathbb{R}^{d \times m}} W_{\text{cont}}(M, t). \end{aligned}$$

This proves $s \in \Theta(M) \cap C$, so $\Theta^-(C)$ is closed and therefore $\Gamma^-(C) = (\nabla y)^{-1}(\Theta^-(C))$ is Lebesgue-measurable. Now, we can apply [FL07, Thm. 6.5], to get a measurable $s \colon \Omega \to \mathbb{R}^{d \times m}$, with

$$W_{\text{cont}}(\nabla y(x), s(x)) = \min_{t \in \mathbb{R}^{d \times m}} W_{\text{cont}}(\nabla y(x), t)$$

and $s \in L^q(\Omega; \mathbb{R}^{d \times m})$, since

$$\begin{aligned} \int_{\Omega} |s(x)|^q \, dx &\leq C \int_{\Omega} W_{\text{cont}}(\nabla y(x), s(x)) + 1 \, dx \\ &= C \int_{\Omega} \min_{s \in \mathbb{R}^{d \times m}} W_{\text{cont}}(\nabla y(x), s) + 1 \, dx \\ &\leq C \int_{\Omega} \min_{s \in \mathbb{R}^{d \times m}} |\nabla y(x)|^p + |s|^q + 1 \, dx \\ &\leq C \int_{\Omega} |\nabla y(x)|^p + 1 \, dx. \end{aligned}$$

It remains to justify the $\Gamma$-convergence result for $F_\varepsilon^{s-\min}$. Suppose $y_k \to y \in W^{1,p}(\Omega; \mathbb{R}^d)$ strongly in $L^p(\Omega; \mathbb{R}^d)$. Choose $s_k \in L^q(\Omega; \mathbb{R}^{d \times m})$

with $F_{\varepsilon_k}^{s-\min}(y_k, \Omega) \leq F_{\varepsilon_k}(y_k, s_k, \Omega) + k^{-1}$. Without loss of generality assuming that $F_{\varepsilon}^{s-\min}(y_k, \Omega)$ is bounded, by passing to a subsequence (not relabeled) we may assume that $s_k \rightharpoonup s_0$ in $L^q$. But then Theorem 1.7 shows that

$$\liminf_{k \to \infty} F_{\varepsilon_k}^{s-\min}(y_k, \Omega) = \liminf_{k \to \infty} F_{\varepsilon_k}(y_k, s_k, \Omega)$$
$$\geq F(y, s_0, \Omega)$$
$$\geq F^{s-\min}(y, \Omega)$$

by the first part of the proof. On the other hand, if $y \in W^{1,p}(\Omega; \mathbb{R}^d)$ is given, choose $s \in L^q(\Omega; \mathbb{R}^{d \times m})$ according to the first part of the proof such that $F^{s-\min}(y, \Omega) = F(y, s, \Omega)$. Then if $(y_k, s_k)$ is a recovery sequence for $(y, s)$ from Theorem 1.7, we obtain

$$\limsup_{k \to \infty} F_{\varepsilon_k}^{s-\min}(y_k, \Omega) \leq \limsup_{k \to \infty} F_{\varepsilon_k}(y_k, s_k, \Omega)$$
$$= F(y, s, \Omega)$$
$$= F^{s-\min}(y, \Omega).$$

$\square$

## Part II

# Existence and Convergence of Solutions to the Atomistic and Continuum Equations of Nonlinear Elasticity Theory

# Chapter 2

# The Models

Let us start with a description of the models. They are, of course, large similarities to the models in the first part. We will still start anew since there are some important differences. This also ensures that the second part can be read independently of the first.

## 2.1  The Continuum Model

Let us start with the continuum model. We consider a bounded, open reference set $\Omega \subset \mathbb{R}^d$, deformations $y\colon \Omega \to \mathbb{R}^d$, a Borel function $W_{\mathrm{cont}}\colon \mathbb{R}^{d\times d} \to (-\infty, \infty]$ which is bounded from below, a body force $f \in L^2(\Omega; \mathbb{R}^d)$, a boundary datum $g \in H^1(\Omega; \mathbb{R}^d)$ and the deformation energy

$$E(y; f) = \int_{\Omega} W_{\mathrm{cont}}(\nabla y(x)) - y(x) f(x) \, dx.$$

First let us look at the static case. Here we want to find local minimizers of this energy (in a suitable topology) constraint to $y$ having boundary values $g$. In a sufficiently regular setting, these are (stable) solutions of the corresponding Euler-Lagrange equations

$$\begin{cases} -\operatorname{div}(DW_{\mathrm{cont}}(\nabla y(x))) &= f(x) \quad \text{in } \Omega, \\ y(x) &= g(x) \quad \text{on } \partial\Omega. \end{cases}$$

In this part of the thesis, we want the assumptions on $W_{\mathrm{cont}}$ to be weak enough to include, e.g., Lennard-Jones-type interactions. Therefore, we should not assume global (quasi-)convexity or growth at infinity and $W_{\mathrm{cont}}$ should be allowed to have singularities. Of course, under such weak assumptions we cannot hope to solve the problem for all $f$

and $g$. Instead, we will look at a stable affine reference deformation $y_{A_0}(x) = A_0 x$ with gradient $A_0 \in \mathbb{R}^{d \times d}$ and show that for all $f$ small enough and all $g$ close enough to the reference deformation there is a unique deformation close to $y_{A_0}$ that solves the problem. Here, stability is yet to be defined.

In the dynamic case, we also have a time interval $[0, T)$, deformations are now maps $y \colon \Omega \times [0, T) \to \mathbb{R}^d$, we have an initial deformation $h_0 \in H^1(\Omega; \mathbb{R}^d)$, initial velocities $h_1 \in L^2(\Omega; \mathbb{R}^d)$, a time dependent body force $f \in L^2(\Omega \times [0, T); \mathbb{R}^d)$, and time dependent Dirichlet boundary data $g \in L^2([0, T); H^1(\Omega; \mathbb{R}^d))$. We now use Newton's second law of motion where the internal forces are given by the first variation of the potential energy. The reference body is assumed to have constant density $\rho$. By choice of units we can just take $\rho \equiv 1$. That means we are looking for (weak) solutions to the initial boundary value problem

$$\begin{cases} \ddot{y}(x,t) - \operatorname{div}(DW_{\mathrm{cont}}(\nabla y(x,t))) &= f(x,t) & \text{in } \Omega \times (0,T), \\ y(x,t) &= g(x,t) & \text{on } \partial\Omega \times (0,T), \\ y(x,0) &= h_0(x) & \text{in } \Omega, \\ \dot{y}(x,0) &= h_1(x) & \text{in } \Omega. \end{cases}$$

## 2.2   The Atomistic Model

Let us first fix some notation. We consider the reference lattice $\varepsilon \mathbb{Z}^d$, where $\varepsilon > 0$ is the lattice spacing. Up to a set of measure zero, we partition $\mathbb{R}^d$ into the cubes $\{z\} + \left(-\frac{\varepsilon}{2}, \frac{\varepsilon}{2}\right)^d$ with $z \in \varepsilon \mathbb{Z}^d$. Given $x \in \mathbb{R}^d$, not in the neglected set of measure zero, we let $\hat{x} \in \varepsilon \mathbb{Z}^d$ be the midpoint of the corresponding cube and $Q_\varepsilon(x)$ the cube itself. Furthermore, for certain symmetry arguments we will later use the point $\bar{x}$ defined as the reflection of $x$ at $\hat{x}$.

Now, atomistic deformations are maps $y \colon \Omega \cap \varepsilon \mathbb{Z}^d \to \mathbb{R}^d$. Again, we will look at the reference configuration $y_{A_0}(x) = A_0 x$, meaning that the reference positions of the atoms are $A_0 \Omega \cap \varepsilon A_0 \mathbb{Z}^d$ in the macroscopic domain $A_0 \Omega$. The deformation energy is supposed to result from local finite range atomic interactions. More precisely, there is a finite set $\mathcal{R} \subset \mathbb{Z}^d \backslash \{0\}$ accounting for the possible interactions, for which we will always assume that $\operatorname{span}_{\mathbb{Z}} \mathcal{R} = \mathbb{Z}^d$ and $\mathcal{R} = -\mathcal{R}$. We then assume that the atoms marked by $x, \tilde{x} \in \varepsilon \mathbb{Z}^d$ can only interact directly if there is a point $z \in \varepsilon \mathbb{Z}^d$ with $x, \tilde{x} \in z + \varepsilon \mathcal{R}$. Furthermore, we assume our system to be translationally invariant such that the interaction can only depend on the matrix of differences $D_{\mathcal{R},\varepsilon} y(x) = (\frac{y(x+\varepsilon\rho)-y(x)}{\varepsilon})_{\rho\in\mathcal{R}}$ with $x \in \varepsilon\mathbb{Z}^d$, where we already use the natural scaling such that $D_{\mathcal{R},\varepsilon} y_{A_0}(x) = (A_0\rho)_{\rho\in\mathcal{R}}$ for all $\varepsilon > 0$. Our site potential $W_{\text{atom}} \colon (\mathbb{R}^d)^{\mathcal{R}} \to (-\infty,\infty]$ is then assumed to be independent of $\varepsilon$. Compare [BLL02] for a detailed discussion of this scaling assumption.

In a lot of situations the atomistic energy might not be given in terms of a site potential $W_{\text{atom}}$ but, e.g. as a sum of pair potentials and angle contributions. Our theory still applies since one can choose a $W_{\text{atom}}$ that represents the same energy by distributing the interaction terms along the atom sites in a sufficiently symmetric way. Even better, their are typically many possible choices of $W_{\text{atom}}$ that are all physically equivalent. In particular, any assumptions on $W_{\text{atom}}$ we make only have to be satisfied by one such choice.

As a mild symmetry assumption on $W_{\text{atom}}$, we will assume throughout that

$$W_{\text{atom}}(A) = W_{\text{atom}}(T(A))$$

for all $A \in (\mathbb{R}^d)^{\mathcal{R}}$, where

$$T(A)_\rho = -A_{-\rho}.$$

This is indeed a quite weak assumption. In a typical situation this just means that we have partitioned the overall energy in such a way, that the site potential is invariant under a point reflection at that atom combined with the natural relabeling.

**Lemma 2.1.** *If $W_{\text{atom}}$ satisfies the symmetry condition and $B \in \mathbb{R}^{d\times d}$, then*

$$D^k W_{\text{atom}}((B\rho)_{\rho\in\mathcal{R}})[T(A_1),\ldots,T(A_k)]$$
$$= D^k W_{\text{atom}}((B\rho)_{\rho\in\mathcal{R}})[A_1,\ldots,A_k]$$

*whenever these derivatives exist.*

*Proof.* This follows directly from $T((B\rho)_{\rho\in\mathcal{R}}) = (B\rho)_{\rho\in\mathcal{R}}$. $\qquad\square$

Letting $R_{\max} = \max\{|\rho| \colon \rho \in \mathcal{R}\}$ and $R_0 = \max\{R_{\max}, \frac{\sqrt{d}}{4}\}$, the discrete gradient $D_{\mathcal{R},\varepsilon} y$ is surely well-defined on the discrete 'semi-interior'

$$\text{sint}_\varepsilon \Omega = \{x \in \Omega \cap \varepsilon\mathbb{Z}^d \colon \text{dist}(x, \partial\Omega) > \varepsilon R_0\}.$$

The total energy below will be defined by a sum over this set, which is justified by our considering variations only on the discrete interior

$$\operatorname{int}_\varepsilon \Omega = \{x \in \Omega \cap \varepsilon \mathbb{Z}^d : \operatorname{dist}(x, \partial\Omega) > 2\varepsilon R_0\},$$

which do not affect the gradients outside the semi-interior, and by prescribing boundary values on the layer $\partial_\varepsilon \Omega = \Omega \cap \varepsilon \mathbb{Z}^d \setminus \operatorname{int}_\varepsilon \Omega$. Now, given a body force $f \colon \varepsilon \mathbb{Z}^d \cap \Omega \to \mathbb{R}^d$ and a boundary datum $g \colon \partial_\varepsilon \Omega \to \mathbb{R}^d$ we define the set of admissible deformations as

$$\mathcal{A}_\varepsilon(\Omega, g) = \{y \colon \Omega \cap \varepsilon \mathbb{Z}^d \to \mathbb{R}^d : y(x) = g(x) \text{ for all } x \in \partial_\varepsilon \Omega\}$$

and the elastic energy of an atomistic deformation $y$ by

$$E_\varepsilon(y; f; g) = \varepsilon^d \sum_{x \in \operatorname{sint}_\varepsilon \Omega} W_{\operatorname{atom}}(D_{\mathcal{R},\varepsilon} y(x)) - \varepsilon^d \sum_{x \in \varepsilon \mathbb{Z}^d \cap \Omega} y(x) f(x)$$

for $y \in \mathcal{A}_\varepsilon(\Omega, g)$ and otherwise $E_\varepsilon(y; f; g) = \infty$.

We remark here that the definition of $R_0$ implies that

$$\Omega_\varepsilon = \bigcup_{z \in \operatorname{int}_\varepsilon \Omega} Q_\varepsilon(z) \subset \Omega$$

which will simplify things later on.

As in the continuous case, our goal is to find local minimizers of the energy which, under suitable assumptions on $W_{\operatorname{atom}}$, are (stable) solutions of the corresponding Euler-Lagrange equation

$$- \operatorname{div}_{\mathcal{R},\varepsilon} \big( DW_{\operatorname{atom}}(D_{\mathcal{R},\varepsilon} y(x)) \big) = f(x),$$

for $x \in \operatorname{int}_\varepsilon \Omega$. Here we write $DW_{\operatorname{atom}}(M) = \big( \frac{\partial W_{\operatorname{atom}}(M)}{\partial M_{i\rho}} \big)_{\substack{1 \le i \le d \\ \rho \in \mathcal{R}}}$ for $M = (M_{i\rho})_{\substack{1 \le i \le d \\ \rho \in \mathcal{R}}} \in \mathbb{R}^{d \times \mathcal{R}} \cong (\mathbb{R}^d)^{\mathcal{R}}$ and

$$\operatorname{div}_{\mathcal{R},\varepsilon} M(x) = \sum_{\rho \in \mathcal{R}} \frac{M_\rho(x) - M_\rho(x - \varepsilon\rho)}{\varepsilon}$$

for any $M \colon \Omega \cap \varepsilon \mathbb{Z}^d \to \mathbb{R}^{d \times \mathcal{R}} \cong (\mathbb{R}^d)^{\mathcal{R}}$. There are, of course, no actual derivatives in space involved here. These are just short notations for the finite difference operators.

There is no reason to hope for existence (or uniqueness) in general. We will also restrict ourselves to 'elastic' solutions that are (macroscopically) sufficiently close to some affine lattice. To find such solutions we will look close to continuous solutions.

For later use, let us define the following discrete norms and seminorms:

$$\|u\|_{\ell^2_\varepsilon(\Lambda)} = \left(\varepsilon^d \sum_{x\in\Lambda} |u(x)|^2\right)^{\frac{1}{2}}$$

for any finite set $\Lambda$ and $u\colon \Lambda \to \mathbb{R}^d$,

$$\|u\|_{h^1_\varepsilon(\mathrm{sint}_\varepsilon \Omega)} = \left(\varepsilon^d \sum_{x\in\mathrm{sint}_\varepsilon \Omega} |D_{\mathcal{R},\varepsilon}u(x)|^2\right)^{\frac{1}{2}}$$

for $u\colon \Omega \cap \varepsilon\mathbb{Z}^d \to \mathbb{R}^d$ and

$$\|u\|_{h^{-1}_\varepsilon(\mathrm{int}_\varepsilon \Omega)} = \sup\left\{\varepsilon^d \sum_{x\in\mathrm{int}_\varepsilon \Omega} u(x)\varphi(x)\colon\right.$$
$$\left.\varphi \in \mathcal{A}_\varepsilon(\Omega,0) \text{ with } \|\varphi\|_{h^1_\varepsilon(\mathrm{sint}_\varepsilon \Omega)} = 1\right\}$$

for $u\colon \mathrm{int}_\varepsilon \Omega \to \mathbb{R}^d$. The $h^1_\varepsilon$-semi-norm is given by the semi-definite symmetric bilinear form

$$(u,v)_{h^1_\varepsilon(\mathrm{sint}_\varepsilon \Omega)} = \varepsilon^d \sum_{x\in\mathrm{sint}_\varepsilon \Omega} D_{\mathcal{R},\varepsilon}u(x)\colon D_{\mathcal{R},\varepsilon}v(x),$$

where $A\colon B = \sum_\rho \sum_j A_{j\rho}B_{j\rho}$. On $\mathcal{A}_\varepsilon(\Omega,0)$ this is a norm and a scalar product, respectively.

Note in particular the norm inequality

$$\sup_{x\in\mathrm{sint}_\varepsilon \Omega} |D_{\mathcal{R},\varepsilon}u(x)| \leq \varepsilon^{-\frac{d}{2}}\|u\|_{h^1_\varepsilon(\mathrm{sint}_\varepsilon \Omega)}$$

which will play a crucial role later on.

Given $g\colon \partial_\varepsilon\Omega \to \mathbb{R}^d$, $y\colon \Omega \cap \varepsilon\mathbb{Z}^d$ minimizes $\|y\|_{h^1_\varepsilon(\mathrm{sint}_\varepsilon \Omega)}$ under the constraint $y(x) = g(x)$ for all $x \in \partial_\varepsilon\Omega$ if and only if $(y,u)_{h^1_\varepsilon(\mathrm{sint}_\varepsilon \Omega)} = 0$ for all $u \in \mathcal{A}_\varepsilon(\Omega,0)$ and $y(x) = g(x)$ for all $x \in \partial_\varepsilon\Omega$. Thus, for every $g\colon \partial_\varepsilon\Omega \to \mathbb{R}^d$ there is precisely one such $y$, it depends linearly on $g$ and is the unique solution to $\mathrm{div}_{\mathcal{R},\varepsilon} D_{\mathcal{R},\varepsilon}y = 0$ with boundary values $g$. We write $y = T_\varepsilon g$. Accordingly, we define the semi-norm

$$\|g\|_{\partial_\varepsilon\Omega} = \|T_\varepsilon g\|_{h^1_\varepsilon(\mathrm{sint}_\varepsilon \Omega)}.$$

We also want to look at the dynamical setting. We then have deformations $y\colon (\Omega \cap \varepsilon\mathbb{Z}^d)\times[0,T] \to \mathbb{R}^d$, body forces $f_{\text{atom}}\colon \text{int}_\varepsilon\,\Omega\times[0,T] \to \mathbb{R}^d$, and boundary conditions $g_{\text{atom}}\colon \partial_\varepsilon\Omega \times [0,T) \to \mathbb{R}^d$. Additionally, we consider an initial configuration $h_{0,\text{atom}}\colon (\Omega \cap \varepsilon\mathbb{Z}^d) \to \mathbb{R}^d$ and initial velocities $h_{1,\text{atom}}\colon (\Omega \cap \varepsilon\mathbb{Z}^d) \to \mathbb{R}^d$ such that the compatibility conditions $h_{0,\text{atom}} \in \mathcal{A}_\varepsilon(\Omega, g_{\text{atom}}(\cdot,0))$ and $h_{1,\text{atom}} \in \mathcal{A}_\varepsilon(\Omega, \dot{g}_{\text{atom}}(\cdot,0))$ hold true. At last, let us assume that all atoms have the same mass $m_\varepsilon = \varepsilon^d \rho$. This scaling ensures that the macroscopic reference body has a finite positive mass density $\rho$. As remarked before, we can assume $\rho = 1$. Note that the scaling of the potential energy and the masses only affects the scaling of time. With our choice the time will not be rescaled and remains a macroscopic quantity. For the body forces, this scaling corresponds to a macroscopic acceleration of each atom (e.g. through gravity).

Again we apply Newton's second law of motion and arrive at the initial boundary value problem

$$\begin{cases} \ddot{y}(x,t) - \text{div}_{\mathcal{R},\varepsilon}\ \big(DW_{\text{atom}}(D_{\mathcal{R},\varepsilon}y(x,t))\big) \\ \qquad\qquad\qquad = f_{\text{atom}}(x,t) & \text{in } \text{int}_\varepsilon\,\Omega \times (0,T), \\ y(x,t) = g_{\text{atom}}(x,t) & \text{on } \partial\Omega_\varepsilon \times [0,T), \\ y(x,0) = h_{0,\text{atom}}(x) & \text{in } \Omega \cap \varepsilon\mathbb{Z}^d, \\ \dot{y}(x,0) = h_{1,\text{atom}}(x) & \text{in } \Omega \cap \varepsilon\mathbb{Z}^d. \end{cases}$$

## 2.3   The Cauchy-Born Rule

As described in detail in the introduction, it is a fundamental problem to identify the correct $W_{\text{cont}}$ that should be taken for the continuous equation so that one can hope for atomistic solutions close by as $\varepsilon$ becomes small enough. The classical ansatz to resolve this question by applying the Cauchy-Born leads to setting $W_{\text{cont}} = W_{\text{CB}}$, where in our setting the Cauchy-Born energy density has the simple mathematical expression

$$W_{\text{CB}}(A) := W_{\text{atom}}((A\rho)_{\rho\in\mathcal{R}}).$$

In the following we will only consider $W_{\text{cont}} = W_{\text{CB}}$, where $W_{\text{atom}}$ is given. Our main goal is to justify this choice rigorously by discussing the behavior of the corresponding solutions.

# Chapter 3

# Stability

A crucial ingredient for our main theorem, but also for further applications, is the concept of atomistic stability. Here we define the continuous and atomistic stability constants, discuss their properties, and give simple characterizations.

## 3.1 Stability Constants

For a bilinear form $L \in \mathbb{R}^{d \times d \times d \times d} \cong \mathrm{Bil}(\mathbb{R}^{d \times d}) \cong L(\mathbb{R}^{d \times d}, \mathbb{R}^{d \times d})$ we will write

$$L[A, B] = \sum_{j,k,l,m=1}^{d} L_{jklm} A_{jk} B_{lm},$$

$$(L[A])_{jk} = \sum_{l,m=1}^{d} L_{jklm} A_{lm}$$

if $A, B \in \mathbb{R}^{d \times d}$, and

$$|L| = \sup\{L[A, B] \colon |A| = |B| = 1\}.$$

Later we will use a similar notation for higher order tensors.

In our problem $L$ is the tensor in the equation

$$-\operatorname{div}(L[\nabla u]) = f,$$

which is the linearization of the continuous equation at the affine deformation $y_{A_0}$ if $L = D^2 W_{\mathrm{CB}}(A_0)$. The condition that ensures existence, uniqueness and regularity and at the same time ensures that solutions are strict local minimizers of the nonlinear energy, and in that sense stability, is the Legendre-Hadamard condition

$$\lambda_{\mathrm{LH}}(L) = \inf_{\xi,\eta\in\mathbb{R}^d\setminus\{0\}} \frac{L[\xi\otimes\eta,\xi\otimes\eta]}{|\xi|^2|\eta|^2} > 0.$$

It is a well known fact, proven by Fourier transformation and a cutoff argument, that

$$\lambda_{\mathrm{LH}}(L) = \inf_{u\in H_0^1(U;\mathbb{R}^d)\setminus\{0\}} \frac{\int_U L[\nabla u(x),\nabla u(x)]\,dx}{\int_U |\nabla u(x)|^2\,dx}$$

for any open, nonempty $U \subset \mathbb{R}^d$. See Theorem C.1 for one direction of the proof. We also introduce a modified version that we will need later:

$$\tilde{\lambda}_{\mathrm{LH}}(L) = \inf_{\xi,\eta\in\mathbb{R}^d\setminus\{0\}} \frac{L[\xi\otimes\eta,\xi\otimes\eta]}{|\xi|^2 \sum_{\rho\in\mathcal{R}} (\rho\eta)^2} > 0.$$

Since $\mathrm{span}_{\mathbb{Z}}\,\mathcal{R} = \mathbb{Z}^d$, we have $\mathrm{span}_{\mathbb{R}}\,\mathcal{R} = \mathbb{R}^d$ and there are $C_1, C_2 > 0$ such that

$$C_1|\eta|^2 \le \sum_{\rho\in\mathcal{R}}(\rho\eta)^2 \le C_2|\eta|^2.$$

Hence,

$$C_1\tilde{\lambda}_{\mathrm{LH}}(L) \le \lambda_{\mathrm{LH}}(L) \le C_2\tilde{\lambda}_{\mathrm{LH}}(L)$$

and, in particular, $\tilde{\lambda}_{\mathrm{LH}}(L) > 0$ if and only if $\lambda_{\mathrm{LH}}(L) > 0$.

In the atomistic setting we instead look at tensors on higher dimensional spaces of the type $K \in \mathbb{R}^{\{1,\dots,d\}\times\mathcal{R}\times\{1,\dots,d\}\times\mathcal{R}} \cong \mathrm{Bil}(\mathbb{R}^{\{1,\dots,d\}\times\mathcal{R}}) \cong L(\mathbb{R}^{\{1,\dots,d\}\times\mathcal{R}}, \mathbb{R}^{\{1,\dots,d\}\times\mathcal{R}})$. Note that with each such $K$ we can associate a tensor of the form $L \in \mathbb{R}^{d\times d\times d\times d}$ by

$$L[A, B] = K[(A\rho)_{\rho\in\mathcal{R}}, (B\rho)_{\rho\in\mathcal{R}}].$$

In our equations we will consider $K = D^2 W_{\mathrm{atom}}((A_0\rho)_{\rho\in\mathcal{R}})$, which then corresponds to $L = D^2 W_{\mathrm{CB}}(A_0)$. It turns out that we need a stronger condition for existence and the local minimizing property in the atomistic case. We define

$$\lambda_\varepsilon(K,\Omega) = \inf_{\substack{y\in\mathcal{A}_\varepsilon(\Omega,0)\\ y\neq 0}} \frac{\varepsilon^d \sum_{x\in\mathrm{sint}_\varepsilon\Omega} K[D_{\mathcal{R},\varepsilon}y(x), D_{\mathcal{R},\varepsilon}y(x)]}{\varepsilon^d \sum_{x\in\mathrm{sint}_\varepsilon\Omega} |D_{\mathcal{R},\varepsilon}y(x)|^2}.$$

Now by atomistic stability we mean that

$$\lambda_{\text{atom}}(K,\Omega) = \inf_{\varepsilon > 0} \lambda_\varepsilon(K,\Omega) > 0.$$

We will first show that $\lambda_{\text{atom}}$ is in fact independent of $\Omega$ and is equivalently given by the minimization of periodic problems. This can be done in the spirit of a thermodynamical limit argument. Let us consider

$$\mathcal{B}_{0,N} = \{y\colon \{0,\ldots,N\}^d \to \mathbb{R}^d\colon y(z) = 0,\ \text{if } \text{dist}(z,\partial(0,N)^d) \leq 2R_0\}$$

and

$$\mathcal{B}_{\text{per},N} = \{y\colon \{0,\ldots,N\}^d \to \mathbb{R}^d\colon y \text{ is } [0,N)^d\text{-periodic.}\}.$$

Whenever necessary we will consider these functions to be periodically extended to $\mathbb{Z}^d$.

Let us define

$$\mu_{0,N} = \inf_{\substack{y \in \mathcal{B}_{0,N} \\ y \neq 0}} \frac{\displaystyle\sum_{x \in \{0,\ldots,N\}^d} K[D_{\mathcal{R},1}y(x), D_{\mathcal{R},1}y(x)]}{\displaystyle\sum_{x \in \{0,\ldots,N\}^d} |D_{\mathcal{R},1}y(x)|^2}$$

and

$$\mu_{\text{per},N} = \inf_{\substack{y \in \mathcal{B}_{\text{per},N} \\ y \text{ not constant}}} \frac{\displaystyle\sum_{x \in \{0,\ldots,N-1\}^d} K[D_{\mathcal{R},1}y(x), D_{\mathcal{R},1}y(x)]}{\displaystyle\sum_{x \in \{0,\ldots,N-1\}^d} |D_{\mathcal{R},1}y(x)|^2}.$$

In this definition we used that $D_{\mathcal{R},1}y(x) = 0$ for all $x \in \mathbb{Z}^d$ implies that $y$ is constant since $\text{span}_{\mathbb{Z}} \mathcal{R} = \mathbb{Z}^d$. Obviously, $\mu_{0,N}$ is non-increasing and $-|K| \leq \mu_{0,N} \leq |K|$ for $N$ sufficiently large. Hence, $\mu_{0,N} \to \mu_0 \in [-|K|,|K|]$, where $\mu_0 = \inf_N \mu_{0,N}$.

**Proposition 3.1.** *We have $\mu_{\text{per},N} \to \mu_0$ as $N \to \infty$ and $\mu_0 = \inf_N \mu_{\text{per},N}$.*

*Proof.* It is clear that $\mu_{\text{per},N} \leq \mu_{0,N}$ for all $N$. Now let $\delta > 0$ and $N \in \mathbb{N}$ with $N \geq 6R_0$. Take a non-constant $y \in \mathcal{B}_{\text{per},N}$ such that

$$\sum_{x \in \{0,\ldots,N-1\}^d} K[D_{\mathcal{R},1}y(x), D_{\mathcal{R},1}y(x)] \leq (\mu_{\text{per},N}+\delta) \sum_{x \in \{0,\ldots,N-1\}^d} |D_{\mathcal{R},1}y(x)|^2.$$

Now we consider $\tilde{y} = \eta y \in \mathcal{B}_{0,MN}$, where $M \in \mathbb{N}$, $M \geq 3$, and $\eta \in C^\infty(\mathbb{R}^d;[0,1])$ such that $\eta(x) = 1$ whenever $\text{dist}(x,((0,MN)^d)^c) \geq$

$4R_0$, $\eta(x) = 0$ whenever $\text{dist}(x, ((0, MN)^d)^c) \leq 2R_0$ and $|\nabla\eta(x)| \leq \frac{1}{R_0}$ for all $x$. A short calculation gives

$$|D_{\mathcal{R},1}\tilde{y}(x)| \leq C(d, |\mathcal{R}|, R_0)\|y\|_\infty$$

for all $x$. Using this we can estimate

$$\sum_{x \in \{0,\ldots,MN-1\}^d} K[D_{\mathcal{R},1}\tilde{y}(x), D_{\mathcal{R},1}\tilde{y}(x)]$$

$$\leq (M-2)^d(\mu_{\text{per},N} + \delta) \sum_{x \in \{0,\ldots,N-1\}^d} |D_{\mathcal{R},1}y(x)|^2$$

$$+ C(d, |\mathcal{R}|, R_0)N^d M^{d-1}|K|\|y\|_\infty^2$$

$$\leq (\mu_{\text{per},N} + \delta) \sum_{x \in \{0,\ldots,MN-1\}^d} |D_{\mathcal{R},1}\tilde{y}(x)|^2$$

$$+ C(d, |\mathcal{R}|, R_0)N^d M^{d-1}(|K| + |\mu_{\text{per},N}| + \delta)\|y\|_\infty^2.$$

But for $M$ large enough we have

$$C(d, |\mathcal{R}|, R_0)N^d M^{d-1}(|K| + |\mu_{\text{per},N}| + \delta)\|y\|_\infty^2$$

$$\leq \delta \sum_{x \in \{0,\ldots,MN-1\}^d} |D_{\mathcal{R},1}\tilde{y}(x)|^2.$$

Therefore,

$$\mu_0 \leq \mu_{0,MN} \leq \mu_{\text{per},N} + 2\delta.$$

The restriction $N \geq 6R_0$ is not problematic when we take the infimum since

$$\mu_{\text{per},jN} \leq \mu_{\text{per},N}$$

for all $j \in \mathbb{N}$.                                                    $\square$

**Proposition 3.2.** *For all open, bounded, nonempty $\Omega \subset \mathbb{R}^d$ we have*

$$\lambda_{\text{atom}}(K, \Omega) = \lim_{\varepsilon \to 0^+} \lambda_\varepsilon(K, \Omega) = \mu_0.$$

*Proof.* Take $z_1, z_2 \in \mathbb{R}^d$ and $0 < a_1 < a_2$ such that

$$\{z_1\} + [0, a_1]^d \subset \Omega \subset \{z_2\} + (0, a_2)^d.$$

Now, define $(z_{1,\varepsilon})_i = \lceil \frac{(z_1)_i}{\varepsilon} \rceil \varepsilon$, $(z_{2,\varepsilon})_i = \lfloor \frac{(z_1)_i}{\varepsilon} \rfloor \varepsilon$, $N_{1,\varepsilon} = \lfloor \frac{a_1}{\varepsilon} \rfloor - 1$ and $N_{2,\varepsilon} = \lceil \frac{a_2}{\varepsilon} \rceil + 1$. Then,

$$z_{1,\varepsilon} + \varepsilon\{0, \ldots, N_{1,\varepsilon}\}^d \subset \Omega \cap \varepsilon\mathbb{Z}^d \subset z_{2,\varepsilon} + \varepsilon\{0, \ldots, N_{2,\varepsilon}\}^d.$$

Given $y \in \mathcal{B}_{0,N_{1,\varepsilon}}$ we can set

$$\tilde{y}(z_{1,\varepsilon} + \varepsilon v) = \varepsilon y(v)$$

for $v \in \{0, \dots, N_{1,\varepsilon}\}^d$ and $\tilde{y}(x) = 0$ else. Then $\tilde{y} \in \mathcal{A}_\varepsilon(\Omega, 0)$,

$$\sum_{x \in \operatorname{sint}_\varepsilon \Omega} K[D_{\mathcal{R},\varepsilon}\tilde{y}(x), D_{\mathcal{R},\varepsilon}\tilde{y}(x)] = \sum_{x \in \{0,\dots,N_{1,\varepsilon}-1\}^d} K[D_{\mathcal{R},1}y(x), D_{\mathcal{R},1}y(x)]$$

and

$$\sum_{x \in \operatorname{sint}_\varepsilon \Omega} |D_{\mathcal{R},\varepsilon}\tilde{y}(x)|^2 = \sum_{x \in \{0,\dots,N_{1,\varepsilon}-1\}^d} |D_{\mathcal{R},1}y(x)|^2.$$

Hence,

$$\mu_{0,\lfloor \frac{a_1}{\varepsilon}\rfloor -1} \geq \lambda_\varepsilon(K, \Omega).$$

Similarly,

$$\mu_{0,\lceil \frac{a_2}{\varepsilon}\rceil +1} \leq \lambda_\varepsilon(K, \Omega).$$

This holds for all $\varepsilon > 0$, if we set $\mu_{0,-1} = \infty$. Therefore,

$$\lim_{\varepsilon \to 0} \lambda_\varepsilon(K, \Omega) = \inf_{\varepsilon > 0} \lambda_\varepsilon(K, \Omega) = \mu_0.$$

$\square$

Because of Proposition 3.2 we can now just write $\lambda_{\text{atom}}(K)$. We will also abuse the notation by writing $\lambda_{\text{LH}}(K)$ for the stability constant of the corresponding $\mathbb{R}^{d \times d \times d \times d}$ tensor. In the case $K = D^2 W_{\text{atom}}((A_0\rho)_{\rho \in \mathcal{R}})$ and $L = D^2 W_{\text{CB}}(A_0)$ with $A_0 \in \mathbb{R}^{d \times d}$ we will also just write $\lambda_{\text{atom}}(A_0)$ and $\lambda_{\text{LH}}(A_0)$ and suppress the $W_{\text{atom}}$ dependency. In the same way we will write $\lambda_{\text{atom}}(A_0)$ instead of $\lambda_{\text{atom}}(D^2 W_{\text{atom}}(A_0))$ if $A_0 \in \mathbb{R}^{d \times \mathcal{R}}$.

For the dependency on $K$ of $\lambda_{\text{atom}}$ we record the following elementary observation.

**Proposition 3.3.** *Given tensors* $K, \tilde{K}$, *we have*

$$|\lambda_{\text{atom}}(K) - \lambda_{\text{atom}}(\tilde{K})| \leq |K - \tilde{K}|.$$

*In particular, if* $W_{\text{atom}} \in C^2(V)$, $V$ *open, then*

$$\{A \in V : \lambda_{\text{atom}}(A) > 0\}$$

*is open as well.*

*Proof.* This is straightforward. Just use

$$\left| \sum_{x \in \{0,\ldots,N\}^d} K[D_{\mathcal{R},1}y(x), D_{\mathcal{R},1}y(x)] - \sum_{x \in \{0,\ldots,N\}^d} \tilde{K}[D_{\mathcal{R},1}y(x), D_{\mathcal{R},1}y(x)] \right|$$

$$\leq |K - \tilde{K}| \sum_{x \in \{0,\ldots,N\}^d} |D_{\mathcal{R},1}y(x)|^2$$

$\square$

## 3.2   Representation Formulae

Combining Proposition 3.1 and Proposition 3.2 we are now basically in the setting of the stability discussion in [HO12]. We include the most important points here to stay self-contained and, more importantly, to provide a new, more intuitive characterization of the stability constant and sufficient criteria for stability that allow for a direct application in interesting situations.

Remark that we use an $h^1$-Norm based on difference quotients, while the authors in [HO12] use a Fourier norm. Therefore, the stability constants will be different, but of course everything remains equivalent. While the Fourier norm makes the connection to the continuum case slightly more direct, our approach has the advantage that it is considerably easier to check if the atomistic stability condition holds true in specific situations making it possible to rigorously discuss relatively simple examples as will be detailed in the Section 3.3.

We will write

$$Q_N = \{0, 1, \ldots, N - 1\}^d$$

and

$$\hat{Q}_N = \left\{0, \frac{2\pi}{N}, \ldots, \frac{2\pi(N-1)}{N}\right\}^d$$

for the dual group. Given $y \colon Q_N \to \mathbb{C}$, its Fourier transformation is defined by $\hat{y} \colon \hat{Q}_N \to \mathbb{C}$ by

$$\hat{y}(k) = \frac{1}{N^d} \sum_{x \in Q_N} y(x) e^{-ixk}.$$

We have

$$\sum_{x \in Q_N} e^{ix(k-k')} = N^d \delta_{k,k'}$$

for all $k, k' \in \hat{Q}_N$ and

$$\sum_{k \in \hat{Q}_N} e^{i(x-x')k} = N^d \delta_{x,x'}$$

for all $x, x' \in Q_N$. Therefore,

$$y(x) = \sum_{k \in \hat{Q}_N} \hat{y}(k) e^{ixk}$$

for all $x \in Q_N$.

In Fourier space the problem is in diagonal form. In the following we will often assume that $K_{j\rho l\sigma} = K_{l\sigma j\rho}$ which is automatically satisfied if $K$ is the second derivative of a potential.

**Proposition 3.4.** *Assume that* $K_{j\rho l\sigma} = K_{l\sigma j\rho}$ *for all* $j, l, \rho, \sigma$. *Now, given* $y \colon Q_N \to \mathbb{R}^d$ *periodically extended to* $\mathbb{Z}^d$, *we have*

$$\sum_{x \in Q_N} K[D_{\mathcal{R},1} y(x), D_{\mathcal{R},1} y(x)] = N^d \sum_{k \in \hat{Q}_N} \hat{y}(k)^T H(k) \overline{\hat{y}(k)},$$

*where*

$$H(k)_{jl} = \sum_{\rho,\sigma \in \mathcal{R}} K_{j\rho l\sigma} \big( \cos(\rho k) - 1 + i \sin(\rho k) \big) \big( \cos(\sigma k) - 1 - i \sin(\sigma k) \big).$$

*In particular,* $H(k)$ *is Hermitian for all* $k$, $H$ *is* $[0, 2\pi)^d$-*periodic and* $\overline{H(k)} = H(-k)$ *for all* $k$.

*Furthermore, if* $K$ *additionally satisfies either* $K_{j\rho l\sigma} = K_{j(-\rho)l(-\sigma)}$ *or* $K_{j\rho l\sigma} = K_{l\rho j\sigma}$, *then*

$$H(k)_{jl} = \sum_{\rho,\sigma \in \mathcal{R}} K_{j\rho l\sigma} \big( (\cos(\rho k) - 1)(\cos(\sigma k) - 1) + \sin(\rho k) \sin(\sigma k) \big).$$

*In particular,* $H(k) \in \mathbb{R}^{d \times d}_{\mathrm{sym}}$ *for all* $k$.

*Remark* 3.5. The condition $K_{j\rho l\sigma} = K_{l\sigma j\rho}$ is automatically satisfied if $K$ is given as a second derivative of a potential. The condition $K_{j\rho l\sigma} = K_{j(-\rho)l(-\sigma)}$ is implied by our mild symmetry condition as shown in Lemma 2.1. The last condition $K_{j\rho l\sigma} = K_{l\rho j\sigma}$ is then not needed and just given as an alternative for the stability analysis. It is connected to the Cauchy relations.

*Proof.*

$$\sum_{x \in Q_N} K[D_{\mathcal{R},1}y(x), D_{\mathcal{R},1}y(x)]$$

$$= \sum_{x,j,l,\rho,\sigma,k,k'} K_{j\rho l\sigma}(e^{i\rho k} - 1)(e^{-i\sigma k'} - 1)e^{ix(k-k')}\hat{y}_j(k)\overline{\hat{y}_l(k')}$$

$$= N^d \sum_{j,l,\rho,\sigma,k} K_{j\rho l\sigma}\big(\cos(\rho k) - 1 + i\sin(\rho k)\big)$$
$$\cdot \big(\cos(\sigma k) - 1 - i\sin(\sigma k)\big)\hat{y}_j(k)\overline{\hat{y}_l(k)}.$$

Everything else follows easily since

$$(\cos(\rho k) - 1 + i\sin(\rho k))(\cos(\sigma k) - 1 - i\sin(\sigma k))$$
$$= (\cos(\rho k) - 1)(\cos(\sigma k) - 1) + \sin(\rho k)\sin(\sigma k)$$
$$+ i(\cos(\sigma k) - 1)\sin(\rho k) - i(\cos(\rho k) - 1)\sin(\sigma k)$$

$$\square$$

**Proposition 3.6.** *Assume that $K_{j\rho l\sigma} = K_{l\sigma j\rho}$ for all $j, l, \rho, \sigma$. Then,*

$$\mu_{per,N} = \min\left\{\frac{h(k)}{\sum_{\rho \in \mathcal{R}}(1 - \cos(\rho k))^2 + \sin^2(\rho k)} : k \in \hat{Q}_N \backslash \{0\}\right\},$$

*where $h(k)$ is the smallest eigenvalue of $H(k)$.*

*Proof.* First of all note that, for $k \in \hat{Q}_N$, $\sum_{\rho \in \mathcal{R}}(1 - \cos(\rho k))^2 + \sin^2(\rho k) = 0$ is equivalent to $k\rho \in 2\pi\mathbb{Z}$ for all $\rho \in \mathcal{R}$ or, equivalently, for all $\rho \in \mathbb{Z}^d$ since $\text{span}_{\mathbb{Z}} \mathcal{R} = \mathbb{Z}^d$. This is the case if and only if $k = 0$. Now set

$$\mu_{\mathrm{F},N} = \min\left\{\frac{h(k)}{\sum_{\rho \in \mathcal{R}}(1 - \cos(\rho k))^2 + \sin^2(\rho k)} : k \in \hat{Q}_N \backslash \{0\}\right\}.$$

Given $y\colon Q_N \to \mathbb{R}^d$, periodically extended, we have

$$
\sum_{x \in Q_N} K[D_{\mathcal{R},1}y(x), D_{\mathcal{R},1}y(x)]
$$

$$
= N^d \sum_{k \in \hat{Q}_N} \hat{y}(k)^T H(k) \overline{\hat{y}(k)}
$$

$$
\geq N^d \sum_{k \in \hat{Q}_N} h(k) |\hat{y}(k)|^2
$$

$$
\geq \mu_{\mathrm{F},N} N^d \sum_{k \in \hat{Q}_N} \sum_{\rho \in \mathcal{R}} |\hat{y}(k)|^2 \big( (1 - \cos(\rho k))^2 + \sin^2(\rho k) \big)
$$

$$
= \mu_{\mathrm{F},N} \sum_{x \in Q_N} |D_{\mathcal{R},1}y(x)|^2,
$$

where we used Proposition 3.4 for $K$ and $\tilde{K}_{j\rho l\sigma} = \delta_{jl}\delta_{\rho\sigma}$. This proves $\mu_{\mathrm{per},N} \geq \mu_{\mathrm{F},N}$. For the opposite inequality take $k_0 \in \hat{Q}_N \backslash \{0\}$ such that

$$
h(k_0) = \mu_{\mathrm{F},N} \sum_{\rho \in \mathcal{R}} (1 - \cos(\rho k))^2 + \sin^2(\rho k).
$$

Let $v_0$ be a corresponding eigenvector and $k_1 \in \hat{Q}_N$ be the unique vector such that $k_0 + k_1 \in 2\pi\mathbb{Z}^d$. In the case $k_0 = k_1$, take $v_0$ real. We define

$$
y(x) = \overline{v}_0 e^{ik_0 x} + v_0 e^{ik_1 x}.
$$

For $x \in Q_N$ we have $y(x) = 2\,\mathrm{Re}\,v_0 \cos(k_0 x) + 2\,\mathrm{Im}\,v_0 \sin(k_0 x)$, which is real, $[0, N)^d$-periodic and non-constant. We calculate

$$
\sum_{x \in Q_N} K[D_{\mathcal{R},1}y(x), D_{\mathcal{R},1}y(x)]
$$

$$
= N^d \sum_{k \in \hat{Q}_N} \hat{y}(k)^T H(k) \overline{\hat{y}(k)}
$$

$$
= 2N^d h(k_0) |v_0|^2 (1 + \delta_{k_0 k_1})
$$

$$
= \mu_{\mathrm{F},N} N^d \sum_{k \in \hat{Q}_N} \sum_{\rho \in \mathcal{R}} |\hat{y}(k)|^2 \big( (1 - \cos(\rho k))^2 + \sin^2(\rho k) \big)
$$

$$
= \mu_{\mathrm{F},N} \sum_{x \in Q_N} |D_{\mathcal{R},1}y(x)|^2.
$$

Therefore, $\mu_{\mathrm{per},N} \leq \mu_{\mathrm{F},N}$. $\qquad \square$

In the limit $N \to \infty$ we get the following result:

**Theorem 3.7.** *Assume that $K_{j\rho l\sigma} = K_{l\sigma j\rho}$ for all $j, l, \rho, \sigma$. Then*

$$\lambda_{\text{atom}}(K) = \inf \left\{ \frac{h(k)}{\sum_{\rho \in \mathcal{R}} (1 - \cos(\rho k))^2 + \sin^2(\rho k)} : k \in [0, 2\pi)^d \backslash \{0\} \right\},$$

$$\tilde{\lambda}_{\text{LH}}(K) = \lim_{s \to 0^+} \inf \left\{ \frac{h(k)}{\sum_{\rho \in \mathcal{R}} (1 - \cos(\rho k))^2 + \sin^2(\rho k)} : \right.$$
$$\left. k \in (-s, s)^d \backslash \{0\} \right\},$$

*where $h(k)$ is the smallest eigenvalue of $H(k)$. In particular, atomistic stability implies the Legendre-Hadamard condition.*

*Proof.* Set

$$\mu_{\text{F}} = \inf \left\{ \frac{h(k)}{\sum_{\rho \in \mathcal{R}} (1 - \cos(\rho k))^2 + \sin^2(\rho k)} : k \in [0, 2\pi)^d \backslash \{0\} \right\}$$

By Proposition 3.6 we have $\mu_{\text{per},N} \geq \mu_{\text{F}}$ and thus $\lambda_{\text{atom}}(K) \geq \mu_{\text{F}}$. For the opposite inequality let $M > \mu_{\text{F}}$. Now, take $k_0 \in [0, 2\pi)^d \backslash \{0\}$ such that

$$\frac{h(k_0)}{\sum_{\rho \in \mathcal{R}} (1 - \cos(\rho k_0))^2 + \sin^2(\rho k_0)} < M.$$

By continuity of $h$, we can find an $N \in \mathbb{N}$ and a $k_1 \in \hat{Q}_N$ such that

$$\frac{h(k_1)}{\sum_{\rho \in \mathcal{R}} (1 - \cos(\rho k_1))^2 + \sin^2(\rho k_1)} < M.$$

Therefore, $\lambda_{\text{atom}}(K) \leq \mu_{\text{per},N} < M$.

Now, let $\eta \in \mathbb{R}^d$ with $|\eta| = 1$ and $0 < \tau \leq 1$. Then,

$$\left| (1 - \cos(\rho \eta \tau))^2 + \sin^2(\rho \eta \tau) - \tau^2 (\rho \eta)^2 \right| \leq C \tau^4$$

and for $\xi \in \mathbb{C}^d$ with $|\xi| = 1$

$$\left| \xi^T H(\eta \tau) \overline{\xi} - \tau^2 K[\xi \otimes (\rho \eta)_{\rho \in \mathcal{R}}, \overline{\xi} \otimes (\rho \eta)_{\rho \in \mathcal{R}}] \right| \leq C \tau^3.$$

This implies

$$\left| h(\eta \tau) - \min_{\substack{\xi \in \mathbb{C}^d \\ |\xi| = 1}} \tau^2 K[\xi \otimes (\rho \eta)_{\rho \in \mathcal{R}}, \overline{\xi} \otimes (\rho \eta)_{\rho \in \mathcal{R}}] \right| \leq C \tau^3.$$

Furthermore, for all $\eta$ as above we have

$$0 < c \leq \sum_{\rho \in \mathcal{R}} (\rho\eta)^2 \leq C$$

and

$$\left| \min_{\substack{\xi \in \mathbb{C}^d \\ |\xi|=1}} K[\xi \otimes (\rho\eta)_{\rho \in \mathcal{R}}, \overline{\xi} \otimes (\rho\eta)_{\rho \in \mathcal{R}}] \right| \leq C.$$

Thus, for $\tau$ small enough we also know that

$$\sum_{\rho \in \mathcal{R}} (1 - \cos(\rho\eta\tau))^2 + \sin^2(\rho\eta\tau) \geq \frac{c\tau^2}{2}.$$

Due to the symmetry of $K$ we have

$$K[\xi \otimes b, \overline{\xi} \otimes b] = K[\operatorname{Re}\xi \otimes b, \operatorname{Re}\xi \otimes b] + K[\operatorname{Im}\xi \otimes b, \operatorname{Im}\xi \otimes b]$$

for all $\xi \in \mathbb{C}^d$ and $b \in \mathbb{R}^{\mathcal{R}}$. In particular,

$$\min_{\substack{\xi \in \mathbb{C}^d \\ |\xi|=1}} K[\xi \otimes (\rho\eta)_{\rho \in \mathcal{R}}, \overline{\xi} \otimes (\rho\eta)_{\rho \in \mathcal{R}}] = \min_{\substack{\xi \in \mathbb{R}^d \\ |\xi|=1}} K[\xi \otimes (\rho\eta)_{\rho \in \mathcal{R}}, \xi \otimes (\rho\eta)_{\rho \in \mathcal{R}}]$$

Combining the above inequalities we get

$$\left| \frac{h(\eta\tau)}{\sum_{\rho \in \mathcal{R}} (1 - \cos(\rho\eta\tau))^2 + \sin^2(\rho\eta\tau)} \right.$$
$$\left. - \frac{\min_{\xi \in \mathbb{R}^d, |\xi|=1} K[\xi \otimes (\rho\eta)_{\rho \in \mathcal{R}}, \xi \otimes (\rho\eta)_{\rho \in \mathcal{R}}]}{\sum_{\rho \in \mathcal{R}} (\rho\eta)^2} \right|$$
$$\leq \frac{4C^2}{c^2} \tau$$

for all $\tau$ small enough and all $\eta$ as above. Therefore,

$$\lim_{\tau \to 0^+} \min_{|\eta|=1} \frac{h(\eta\tau)}{\sum_{\rho \in \mathcal{R}} (1 - \cos(\rho\eta\tau))^2 + \sin^2(\rho\eta\tau)} = \tilde{\lambda}_{\mathrm{LH}}(K)$$

which gives the desired result. $\qquad\square$

If $H$ is real we can express $\lambda_{\mathrm{atom}}$ in a way that looks quite similar to the definition of $\lambda_{\mathrm{LH}}$.

**Corollary 3.8.** *Assume that we have* $K_{j\rho l\sigma} = K_{l\sigma j\rho}$ *and additionally that* $K_{j\rho l\sigma} = K_{j(-\rho)l(-\sigma)}$ *or* $K_{j\rho l\sigma} = K_{l\rho j\sigma}$ *for all* $j, l, \rho, \sigma$. *Then*

$$\lambda_{\text{atom}}(K) = \inf \left\{ \frac{K[\xi \otimes c(k), \xi \otimes c(k)] + K[\xi \otimes s(k), \xi \otimes s(k)]}{|\xi|^2(|c(k)|^2 + |s(k)|^2)} : \right.$$
$$\left. \xi \in \mathbb{R}^d \backslash \{0\}, k \in [0, 2\pi)^d \backslash \{0\} \right\},$$

*where* $c(k)_\rho = \cos(\rho k) - 1$ *and* $s(k)_\rho = \sin(\rho k)$.

The following criterion is strictly weaker but often easier to check.

**Corollary 3.9.** *Assume that we have* $K_{j\rho l\sigma} = K_{l\sigma j\rho}$ *and additionally that* $K_{j\rho l\sigma} = K_{j(-\rho)l(-\sigma)}$ *or* $K_{j\rho l\sigma} = K_{l\rho j\sigma}$ *for all* $j, l, \rho, \sigma$. *Let* $\lambda_{\text{LH}}(K) > 0$, $K[\xi \otimes s(k), \xi \otimes s(k)] \geq 0$ *for all* $\xi, k \in \mathbb{R}^d$ *and*

$$K[\xi \otimes c(k), \xi \otimes c(k)] \geq \gamma |\xi|^2 |c(k)|^2$$

*for all* $\xi, k \in \mathbb{R}^d$ *and some* $\gamma > 0$.
*Then* $\lambda_{\text{atom}}(K) > 0$.

*Proof.* Since $\lambda_{\text{LH}}(K)$ and $\tilde{\lambda}_{\text{LH}}(K)$ are equivalent, we can use Theorem 3.7 to see that that there are some $\tilde{\gamma}, \delta > 0$ such that

$$K[\xi \otimes c(k), \xi \otimes c(k)] + K[\xi \otimes s(k), \xi \otimes s(k)] \geq \tilde{\gamma} |\xi|^2 (|c(k)|^2 + |s(k)|^2)$$

for all $\xi$ and all $k$ with $\text{dist}(k, 2\pi\mathbb{Z}^d) < \delta$. On the other hand, there is a $C > 0$ such that $|s(k)| \leq C|c(k)|$ whenever $\text{dist}(k, 2\pi\mathbb{Z}^d) \geq \delta$. Therefore

$$K[\xi \otimes c(k), \xi \otimes c(k)] + K[\xi \otimes s(k), \xi \otimes s(k)] \geq \frac{\gamma}{1 + C^2} |\xi|^2 (|c(k)|^2 + |s(k)|^2)$$

for these $k$ and all $\xi$. □

*Remark* 3.10. The connection to the characterization in [HO12] is given by the formulas

$$4 \sin^2\left(\frac{z}{2}\right) = (\cos(z) - 1)^2 + \sin^2(z)$$

and

$$2 \sin^2\left(\frac{y}{2}\right) + 2 \sin^2\left(\frac{z}{2}\right) - 2 \sin^2\left(\frac{z - y}{2}\right)$$
$$= (\cos(y) - 1)(\cos(z) - 1) + \sin(y)\sin(z).$$

A little bit of calculation shows that the stability constants here and in [HO12] then are actually equivalent (with the minor correction, that most of their sums should actually run over the set $\mathcal{R} - \mathcal{R}$ instead of $\mathcal{R}$).

## 3.3 Examples

First of all, it is important to point out that the general assumptions made in this part of the thesis are consistent with a large variety of atomic interaction models and lattices. A simple sufficient condition for atomistic stability is the following:

**Proposition 3.11.** *If $W_{\mathrm{atom}} \in C^2$ in a neighborhood of $(A_0\rho)_{\rho\in\mathcal{R}}$, satisfies the symmetry condition, and $(A_0\rho)_{\rho\in\mathcal{R}}$ is a local minimizer of $W_{\mathrm{atom}}$, such that the second derivative in the directions of affine rank-one deformations $((\xi \otimes \eta)\rho)_{\rho\in\mathcal{R}}$ and on the orthogonal complement of all affine deformations is strictly positive. Then $\lambda_{\mathrm{atom}}(A_0) > 0$.*

*Proof.* Just use Corollary 3.9 and the fact that $\xi \otimes c(k)$ is orthogonal on affine deformations. $\qquad\qquad\square$

*Remark* 3.12. These conditions allow for a large class of frame indifferent interaction models. Examples include the general finite range potentials discussed in [CDKM06].

We next want to discuss the connection between atomistic and continuous stability. While atomistic stability always implies the Legendre-Hadamard condition, it is a much harder problem to understand under what condition they are equivalent or strictly different. Indeed the current conjecture seems to be that both cases can happen generically, compare [HO12]. So in certain regimes one has $\lambda_{\mathrm{atom}}(A) = \tilde{\lambda}_{\mathrm{LH}}(A)$ for a large set of matrices or at least $\lambda_{\mathrm{atom}}(A) > 0$ if and only if $\lambda_{\mathrm{LH}}(A) > 0$. While in other regimes or for different potentials it can indeed happen that $\lambda_{\mathrm{atom}}(A) < 0 < \lambda_{\mathrm{LH}}(A)$. In [HO12] this is indeed discussed analytically in the one-dimensional case and numerically in higher dimensions. In the following, the main aim is therefore to give rigorous and multi-dimensional examples for both cases. The examples are two-dimensional to allow for a significantly easier analytical treatment, but the studied effects are expected to be the same in three dimensions.

## An Example on the Triangular Lattice

First, let us look at a rather simple but multidimensional example where it is possible to analytically prove $\lambda_{\text{atom}}(A) = \tilde{\lambda}_{\text{LH}}(A)$ for a large set of matrices $A$. To be more precise, we consider uniform contractions and extensions of a triangular lattice where the energy is given by an unspecified pair potential for the nearest neighbors. This means we will look at $d = 2$,

$$M = \begin{pmatrix} 1 & \frac{1}{2} \\ 0 & \frac{\sqrt{3}}{2} \end{pmatrix}$$

and consider the linearization at $M(t) = tM$ for $t > 0$. Furthermore, $\mathcal{R} = \{\pm e_1, \pm e_2, \pm(e_2 - e_1)\}$ and the interaction is given by

$$W_{\text{atom}}(A) = \frac{1}{2} \sum_{\rho \in \mathcal{R}} V_0(|A\rho|)$$

with some pair potential $V_0 \in C^2((0, \infty); \mathbb{R})$. The Cauchy-Born energy density is then given by

$$W_{\text{CB}}(A) = V_0(|A_{\cdot 1}|) + V_0(|A_{\cdot 2}|) + V_0(|A_{\cdot 2} - A_{\cdot 1}|).$$

Direct calculations give

$$K(t)_{j\rho l\sigma} = \delta_{\rho\sigma} \left( \frac{V_0'(t)}{t} (\delta_{jl} - (M\rho)_j (M\rho)_l) + V_0''(t)(M\rho)_j(M\rho)_l \right)$$

and

$$
\begin{aligned}
H(t, k) &= \\
&= 8 \frac{V_0'(t)}{t} \left( \sin^2\left(\frac{k_1}{2}\right) + \sin^2\left(\frac{k_2}{2}\right) + \sin^2\left(\frac{k_2 - k_1}{2}\right) \right) \text{Id} \\
&\quad + 2\left( V_0''(t) - \frac{V_0'(t)}{t} \right) \left( \sin^2\left(\frac{k_1}{2}\right) \begin{pmatrix} 4 & 0 \\ 0 & 0 \end{pmatrix} + \sin^2\left(\frac{k_2}{2}\right) \begin{pmatrix} 1 & \sqrt{3} \\ \sqrt{3} & 3 \end{pmatrix} \right. \\
&\quad \left. + \sin^2\left(\frac{k_2 - k_1}{2}\right) \begin{pmatrix} 1 & -\sqrt{3} \\ -\sqrt{3} & 3 \end{pmatrix} \right) \\
&= 4\left( V_0''(t) + \frac{V_0'(t)}{t} \right) \left( \sin^2\left(\frac{k_1}{2}\right) + \sin^2\left(\frac{k_2}{2}\right) + \sin^2\left(\frac{k_2 - k_1}{2}\right) \right) \text{Id} \\
&\quad + 2\left( V_0''(t) - \frac{V_0'(t)}{t} \right) \left( \sin^2\left(\frac{k_1}{2}\right) \begin{pmatrix} 2 & 0 \\ 0 & -2 \end{pmatrix} + \sin^2\left(\frac{k_2}{2}\right) \begin{pmatrix} -1 & \sqrt{3} \\ \sqrt{3} & 1 \end{pmatrix} \right. \\
&\quad \left. + \sin^2\left(\frac{k_2 - k_1}{2}\right) \begin{pmatrix} -1 & -\sqrt{3} \\ -\sqrt{3} & 1 \end{pmatrix} \right).
\end{aligned}
$$

The smallest eigenvalue of a matrix

$$\begin{pmatrix} a+b & c \\ c & a-b \end{pmatrix}$$

is $a - \sqrt{b^2 + c^2}$. Hence,

$$
\begin{aligned}
h(t,k) = {}& 4\left(V_0''(t) + \frac{V_0'(t)}{t}\right)\left(\sin^2\left(\frac{k_1}{2}\right) + \sin^2\left(\frac{k_2}{2}\right) + \sin^2\left(\frac{k_2 - k_1}{2}\right)\right) \\
& - 2\sqrt{2}\left|V_0''(t) - \frac{V_0'(t)}{t}\right|\left(\left(\sin^2\left(\frac{k_1}{2}\right) - \sin^2\left(\frac{k_2}{2}\right)\right)^2\right. \\
& + \left(\sin^2\left(\frac{k_1}{2}\right) - \sin^2\left(\frac{k_2 - k_1}{2}\right)\right)^2 \\
& + \left.\left(\sin^2\left(\frac{k_2 - k_1}{2}\right) - \sin^2\left(\frac{k_2}{2}\right)\right)^2\right)^{\frac{1}{2}}.
\end{aligned}
$$

Defining

$$
\begin{aligned}
f(k) = {}& \left(\left(\sin^2\left(\frac{k_1}{2}\right) - \sin^2\left(\frac{k_2}{2}\right)\right)^2 + \left(\sin^2\left(\frac{k_1}{2}\right) - \sin^2\left(\frac{k_2 - k_1}{2}\right)\right)^2\right. \\
& + \left.\left(\sin^2\left(\frac{k_2 - k_1}{2}\right) - \sin^2\left(\frac{k_2}{2}\right)\right)^2\right)^{\frac{1}{2}} \\
& \cdot \left(\sin^2\left(\frac{k_1}{2}\right) + \sin^2\left(\frac{k_2}{2}\right) + \sin^2\left(\frac{k_2 - k_1}{2}\right)\right)^{-1},
\end{aligned}
$$

we have

$$
\begin{aligned}
\lambda_{\mathrm{atom}}(M(t)) = {}& \frac{1}{2}\left(V_0''(t) + \frac{V_0'(t)}{t}\right) \\
& - \frac{1}{2\sqrt{2}}\left|V_0''(t) - \frac{V_0'(t)}{t}\right| \sup_{k \in [0, 2\pi)^d \setminus \{0\}} f(k)
\end{aligned}
$$

and

$$
\begin{aligned}
\tilde{\lambda}_{\mathrm{LH}}(M(t)) = {}& \frac{1}{2}\left(V_0''(t) + \frac{V_0'(t)}{t}\right) \\
& - \frac{1}{2\sqrt{2}}\left|V_0''(t) - \frac{V_0'(t)}{t}\right| \lim_{s \to 0^+} \sup_{k \in (-s,s)^d \setminus \{0\}} f(k).
\end{aligned}
$$

Actually, we claim that

$$
\lambda_{\mathrm{atom}}(M(t)) = \tilde{\lambda}_{\mathrm{LH}}(M(t)) = \frac{1}{2}\left(V_0''(t) + \frac{V_0'(t)}{t}\right) - \frac{1}{4}\left|V_0''(t) - \frac{V_0'(t)}{t}\right|,
$$

which follows from the following lemma.

**Lemma 3.13.** *We have*

$$\lim_{s \to 0^+} \sup_{k \in (-s,s)^2 \setminus \{0\}} f(k) = \sup_{k \in [0,2\pi)^2 \setminus \{0\}} f(k) = \frac{1}{\sqrt{2}}.$$

*Proof.* Looking at $k = (2\tau, 0)$ we have

$$\lim_{\tau \to 0} f(2\tau, 0)^2 = \lim_{\tau \to 0} \frac{2 \sin^4(\tau)}{4 \sin^4(\tau)} = \frac{1}{2}.$$

Hence,

$$\frac{1}{\sqrt{2}} \leq \lim_{s \to 0^+} \sup_{k \in (-s,s)^2 \setminus \{0\}} f(k) \leq \sup_{k \in [0,2\pi)^2 \setminus \{0\}} f(k).$$

The opposite inequality is much more difficult. By periodicity we can look at $k \in [-\pi, \pi]^2 \setminus \{0\}$. We substitute

$$(a, b) = \left( \sin \left( \frac{k_1}{2} \right), \sin \left( \frac{k_2}{2} \right) \right) \in [-1, 1]^2 \setminus \{0\}.$$

Using $\cos \left( \frac{k_1}{2} \right), \cos \left( \frac{k_2}{2} \right) \geq 0$, we can write

$$\sin^2 \left( \frac{k_2 - k_1}{2} \right) = a^2 + b^2 - 2a^2 b^2 - 2ab \sqrt{1 - a^2} \sqrt{1 - b^2}.$$

Therefore, it is sufficient to show the algebraic inequality

$$(a^2 - b^2)^2 + \left( b^2 - 2a^2 b^2 - 2ab \sqrt{1 - a^2} \sqrt{1 - b^2} \right)^2$$
$$+ \left( a^2 - 2a^2 b^2 - 2ab \sqrt{1 - a^2} \sqrt{1 - b^2} \right)^2$$
$$\leq \frac{1}{2} \left( 2a^2 + 2b^2 - 2a^2 b^2 - 2ab \sqrt{1 - a^2} \sqrt{1 - b^2} \right)^2$$

for all $(a, b) \in [-1, 1]^2 \setminus \{0\}$. Evaluating the squares and canceling the terms that appear on both sides, we get the equivalent condition

$$8a^4 b^4 + 8a^3 b^3 \sqrt{1 - a^2} \sqrt{1 - b^2} \leq 4a^2 b^2 (a^2 + b^2)$$

for all $(a, b) \in [-1, 1]^2 \setminus \{0\}$ which can be simplified to

$$2a^2 b^2 + 2ab \sqrt{1 - a^2} \sqrt{1 - b^2} \leq a^2 + b^2$$

for all $a, b \in (0, 1)$ or

$$4a^2 b^2 (1 - a^2)(1 - b^2) \leq (a^2 + b^2 - 2a^2 b^2)^2$$

for all $a, b \in (0, 1)$. Again evaluating the squares and canceling the terms that appear on both sides we finally arrive at the equivalent condition

$$(a^2 - b^2)^2 \geq 0$$

for all $a, b \in (0, 1)$. Furthermore, the proof also shows that $f(k) = \frac{1}{\sqrt{2}}$ if and only if $\sin\left(\frac{k_1}{2}\right) = 0$, $\sin\left(\frac{k_2}{2}\right) = 0$ or $\sin\left(\frac{k_1}{2}\right) = \sin\left(\frac{k_2}{2}\right)$.            $\square$

We have proven the following proposition:

**Proposition 3.14.** *In this setting discussed above we have*

$$\lambda_{\mathrm{atom}}(M(t)) = \tilde{\lambda}_{\mathrm{LH}}(M(t)) = \frac{1}{2}\left(V_0''(t) + \frac{V_0'(t)}{t}\right) - \frac{1}{4}\left|V_0''(t) - \frac{V_0'(t)}{t}\right|.$$

*Remark* 3.15. If $V_0$ is a standard Lennard-Jones potential, i.e.,

$$V_0(r) = r^{-12} - 2r^{-6},$$

$M(t)$ is stable if and only if

$$t \in \left(0, \sqrt[6]{\frac{19}{10}}\right)$$

and $\sqrt[6]{\frac{19}{10}} \approx 1.113$.

*Remark* 3.16. In the proof the choice of our $h^1$-norm helped us to drastically simplify the problem. If one tries to show the equivalent result for the Fourier-$h^1$-norm one has to prove a fully nonlinear, nonalgebraic inequality while in the proof above only a few algebraic manipulations were necessary. Furthermore, even then equality of the two stability constants coming from the Fouriernorm only holds while they are negative.

## An Example on the Rectangular Lattice

Another simple example to illustrate the different notions of stability is the rectangular lattice with nearest and next-to-nearest neighbor interactions that are not balanced with each other. In [FT02] the Cauchy-Born rule is discussed for this problem, which is closely related to our concept of stability. We will discuss precisely their example: Let us set $\mathcal{R} = \{\pm e_1, \pm e_2, e_1 \pm e_2, -e_1 \pm e_2\}$ and

$$W_{\mathrm{atom}}(A) = \frac{K_1}{4} \sum_{\rho \in \mathcal{R}, |\rho|=1} (|A_\rho| - a_1)^2 + \frac{K_2}{4} \sum_{\rho \in \mathcal{R}, |\rho|=\sqrt{2}} (|A_\rho| - a_2)^2$$

for some $a_1, a_2, K_1, K_2 > 0$. We are now interested in the stability of $A_0 = r^*\,\mathrm{Id}$ with

$$r^* = \frac{K_1 a_1 + \sqrt{2} K_2 a_2}{K_1 + 2K_2}.$$

This choice of $r^*$ is not needed to do any analysis, but simplifies the problem since it is the unique $r > 0$ with $DW_{\mathrm{CB}}(r\,\mathrm{Id}) = 0$. Remark though that $DW_{\mathrm{atom}}((r^*\rho)_{\rho\in\mathcal{R}}) \neq 0$ unless $\sqrt{2}a_1 = a_2$. In the following let us use the notation

$$\alpha = \frac{a_2}{\sqrt{2}a_1}, \quad \kappa = \frac{K_2}{K_1} \quad \text{and } \beta = \frac{1 + 2\kappa}{1 + 2\alpha\kappa}.$$

**Proposition 3.17.** *In this setting we have*

$$\tilde{\lambda}_{\mathrm{LH}}(r^*\,\mathrm{Id}) = \frac{K_1}{12}\beta\min\{1, 2\alpha\kappa\} > 0$$

*for all parameter values, while $\lambda_{\mathrm{atom}}(r^*\,\mathrm{Id}) > 0$ if and only if $\beta < 2$, which corresponds to $\alpha \geq \frac{1}{2}$ or $\alpha < \frac{1}{2}$ and $\kappa < \frac{1}{2(1-2\alpha)}$.*

*Proof.* We easily calculate

$$DW_{\mathrm{atom}}(A)[B] = \frac{K_1}{2}\sum_{\rho\in\mathcal{R},|\rho|=1}(|A_\rho| - a_1)\frac{A_\rho \cdot B_\rho}{|A_\rho|}$$
$$+ \frac{K_2}{2}\sum_{\rho\in\mathcal{R},|\rho|=\sqrt{2}}(|A_\rho| - a_2)\frac{A_\rho \cdot B_\rho}{|A_\rho|}$$

and

$$K_{j\rho l\sigma} = D^2 W_{\mathrm{atom}}((r^*\rho)_{\rho\in\mathcal{R}})[e_j \otimes e_\rho, e_l \otimes e_\sigma]$$
$$= \delta_{\rho\sigma}\delta_{|\rho|1}\Big(\delta_{jl}\frac{K_1}{2}\big(1 - \frac{a_1}{r^*}\big) + \rho_j\rho_l\frac{K_1 a_1}{2r^*}\Big)$$
$$+ \delta_{\rho\sigma}\delta_{|\rho|\sqrt{2}}\Big(\delta_{jl}\frac{K_2}{2}\big(1 - \frac{a_2}{\sqrt{2}r^*}\big) + \rho_j\rho_l\frac{K_2 a_2}{4\sqrt{2}r^*}\Big).$$

This gives us

$$H(k) = \begin{pmatrix} a(k) + b(k) & c(k) \\ c(k) & a(k) - b(k) \end{pmatrix},$$

where

$$a(k) = 4K_1\Big(\sin^2\big(\frac{k_1}{2}\big) + \sin^2\big(\frac{k_2}{2}\big)\Big)\Big(1 - \frac{\beta}{2}\Big)$$
$$+ 4K_1\kappa\Big(\sin^2\big(\frac{k_1+k_2}{2}\big) + \sin^2\big(\frac{k_1-k_2}{2}\big)\Big)\Big(1 - \frac{\alpha\beta}{2}\Big),$$
$$b(k) = 2K_1\beta\Big(\sin^2\big(\frac{k_1}{2}\big) - \sin^2\big(\frac{k_2}{2}\big)\Big),$$
$$c(k) = 2K_1\kappa\alpha\beta\Big(\sin^2\big(\frac{k_1+k_2}{2}\big) - \sin^2\big(\frac{k_1-k_2}{2}\big)\Big).$$

In this notation we have

$$h(k) = a(k) - \sqrt{b(k)^2 + c(k)^2}.$$

Substituting $t = \sin(k_1)$ and $s = \sin(k_2)$ we get

$$\lambda_{\text{atom}} = K_1 \inf_{(s,t)\in[-1,1]^2\backslash\{0\}} \Bigg(\frac{(2\kappa+1)(s^2+t^2) - 4\kappa(2-\alpha\beta)s^2t^2}{12(s^2+t^2) - 16s^2t^2}$$
$$- \frac{\sqrt{\beta^2(s^2-t^2)^2 + 16\kappa^2\alpha^2\beta^2 s^2 t^2(1-s^2)(1-t^2)}}{12(s^2+t^2) - 16s^2t^2}\Bigg)$$

In the limit $(s,t) \to 0$ this shows

$$\tilde{\lambda}_{\text{LH}} = \frac{K_1}{12}\beta \min\{1, 2\alpha\kappa\} > 0.$$

For $\beta \geq 2$, we just have to test with $s = t = 1$ to find

$$\lambda_{\text{atom}} \leq \frac{K_1\big(2(2\kappa+1) - 4\kappa(2-\alpha\beta)\big)}{8} = \frac{K_1(2-\beta)}{4} \leq 0.$$

Let us discuss the case $\beta < 2$. First note that

$$2 - \beta = 2\kappa\alpha\beta + 1 - 2\kappa$$

so that $\beta < 2$ is equivalent to $2\kappa\alpha\beta + 1 > 2\kappa$. But then we have

$$(2\kappa+1)(\frac{1}{s^2} + \frac{1}{t^2}) > 4\kappa(2-\alpha\beta)$$

for all $s, t \in (0,1)$. Therefore,

$$(2\kappa+1)(s^2+t^2) - 4\kappa(2-\alpha\beta)s^2t^2 > 0$$

for all $s, t \in [-1, 1]^2 \backslash \{0\}$. Since we have already discussed the limit $(s, t) \to 0$ atomistic stability is now equivalent to

$$\left((2\kappa + 1)(s^2 + t^2) - 4\kappa(2 - \alpha\beta)s^2 t^2\right)^2$$
$$> \beta^2 (s^2 - t^2)^2 + 16\kappa^2 \alpha^2 \beta^2 s^2 t^2 (1 - s^2)(1 - t^2)$$

for all $s, t \in [-1, 1]^2 \backslash \{0\}$. If either $s = 0$ or $t = 0$ this inequality is always true since $\beta < 2\kappa + 1$. Hence, we can substitute $a = \frac{1}{s^2} - 1$, $b = \frac{1}{t^2} - 1$ and divide by $s^4 t^4$ to get the equivalent condition

$$\left((2\kappa + 1)(a + b) + 2(2\kappa + 1) - 4\kappa(2 - \alpha\beta)\right)^2 > \beta^2(b - a)^2 + 16\kappa^2 \alpha^2 \beta^2 ab$$

for all $a, b \geq 0$. I.e.,

$$(a^2 + b^2)\left((2\kappa + 1)^2 - \beta^2\right) + 2ab\left((2\kappa + 1)^2 + \beta^2 - 8\kappa^2 \alpha^2 \beta^2\right)$$
$$+ (a + b)2(2\kappa + 1)(2 - 4\kappa + 4\kappa\alpha\beta) + (2 - 4\kappa + 4\kappa\alpha\beta)^2$$
$$> 0$$

for all $a, b \geq 0$. Simplifying the terms we get the condition

$$(a^2 + b^2)\beta^2\left((2\kappa\alpha + 1)^2 - 1\right) + 2ab\beta^2\left((2\kappa\alpha + 1)^2 + 1 - 8\kappa^2 \alpha^2\right)$$
$$+ (a + b)4(2\kappa + 1)(2 - \beta) + 4(2 - \beta)^2$$
$$> 0$$

which turns out to be just

$$(a - b)^2 4\kappa^2 \alpha^2 \beta^2 + (a + b)^2 4\kappa\alpha\beta^2 + 4ab\beta^2 + 4(a + b)(2\kappa + 1)(2 - \beta)$$
$$+ 4(2 - \beta)^2 > 0$$

for all $a, b \geq 0$ which is obviously true.                                        $\square$

In this example we see that the Legendre-Hadamard stability constant and the atomistic stability constant can be quite different and the parameter regions where we have macroscopic or atomistic stability can be very different as well. In our Fourier characterization it is clear that this difference occurs whenever a system is stable under macroscopic, long wavelength perturbations but not under some perturbation with wavelength on a smaller scale. In this example, the instability occurred on the atomistic scale and actually corresponds to a period doubling (remember that $t = s = 1$ corresponds to $k = (\pi, \pi)$).

The example is actually much more general than it looks. Given general pair potentials $V_1, V_2 \in C^2(0, \infty)$ as well as an $r^*$ with

$$V_1(r^*) + \sqrt{2}V_2'(r^*) = 0,$$

one can look at the site potential

$$W_{\text{atom}}(A) = \frac{1}{2} \sum_{\rho \in \mathcal{R}, |\rho|=1} V_1(|A_\rho|) + \frac{1}{2} \sum_{\rho \in \mathcal{R}, |\rho|=\sqrt{2}} V_2(|A_\rho|).$$

We can now set $K_1 = V_1''(r^*)$, $K_2 = V_2''(r^*)$, $a_1 = r^* - \frac{V_1'(r^*)}{V_1''(r^*)}$, and $a_2 = \sqrt{2}r^* - \frac{V_2'(\sqrt{2}r^*)}{V_2''(\sqrt{2}r^*)}$. As long as $K_1, K_2, a_1, a_2 > 0$, the above analysis applies directly since the linearization $K$ is the same.

# Chapter 4

# The Static Case

## 4.1 The Continuum Problem

### The Linearized System

Let us first recall existing results for the linear(-ized) system.

**Theorem 4.1.** *Let $\Omega \subset \mathbb{R}^d$ be an open, bounded set, let $L \in \mathbb{R}^{d \times d \times d \times d}$, $f \in L^2(\Omega; \mathbb{R}^d)$, $F \in L^2(\Omega; \mathbb{R}^{d \times d})$ and $g \in H^1(\Omega; \mathbb{R}^d)$. Furthermore, assume $\lambda_{\mathrm{LH}}(L) > 0$. Then there is one and only one weak solution $u \in g + H_0^1(\Omega; \mathbb{R}^d)$ of*

$$- \operatorname{div}(L[\nabla u]) = f - \operatorname{div} F.$$

**Theorem 4.2.** *Let $m \in \mathbb{N}_0$, let $\Omega \subset \mathbb{R}^d$ be an open, bounded set with $C^{m+2}$ boundary, let $L \in C^{m,1}(\Omega; \mathbb{R}^{d \times d \times d \times d})$, $f \in H^m(\Omega; \mathbb{R}^d)$, $F \in H^{m+1}(\Omega; \mathbb{R}^{d \times d})$ and $g \in H^{m+2}(\Omega; \mathbb{R}^d)$. Furthermore, assume $\lambda_{\mathrm{LH}}(L(x)) \geq \lambda_0$ for some $\lambda_0 > 0$ and all $x \in \Omega$. Assume that $u \in g + H_0^1(\Omega; \mathbb{R}^d)$ is a weak solution of*

$$- \operatorname{div}(L[\nabla u]) = f - \operatorname{div} F.$$

*Then $u \in H^{m+2}(\Omega; \mathbb{R}^d)$ and there is a $c = c(m, \Omega, \|L\|_{C^{m,1}}, \lambda) > 0$, such that*

$$\|\nabla^{m+2} u\|_{L^2} \leq c(\|f\|_{H^m} + \|\nabla F\|_{H^m} + \|g\|_{H^{m+2}}).$$

We only need this theorem for constant $L$. Reformulating these results we get:

97

**Corollary 4.3.** *Let $m \in \mathbb{N}_0$, let $\Omega \subset \mathbb{R}^d$ be an open, bounded set with $C^{m+2}$ boundary, let $L \in \mathbb{R}^{d \times d \times d \times d}$ and assume $\lambda_{\mathrm{LH}}(L) > 0$. Then the mapping*

$$u \mapsto \mathrm{div}(L[\nabla u])$$

*is a linear isomorphism from $H^{m+2}(\Omega; \mathbb{R}^d) \cap H_0^1(\Omega; \mathbb{R}^d)$ onto $H^m(\Omega; \mathbb{R}^d)$.*

*Proof.* These statements are rather standard and can be found in the literature. See, e.g., [GM12, Corollary 3.46] and [GM12, Theorem 4.14]. In Appendix C we will actually prove more general results for less regular coefficients and general integrability exponent $1 < p < \infty$. □

## Local Solutions of the Nonlinear Problem

Now, let us improve the linearized result to a local result with an implicit function theorem.

**Theorem 4.4.** *Let $m \in \mathbb{N}_0$, $d < 2m + 2$ and let $\Omega \subset \mathbb{R}^d$ be an open, bounded set with $C^{m+2}$-boundary. Furthermore, let $r_0 > 0$, $W_{\mathrm{atom}} \in C^{m+3}\big(\overline{B_{r_0}((A_0\rho)_{\rho \in \mathcal{R}})}\big)$ and assume that $\lambda_{\mathrm{LH}}(A_0) > 0$. Then there are constants $\kappa_1, \kappa_2 > 0$ such that for every $g \in H^{m+2}(\Omega; \mathbb{R}^d)$ and every $f \in H^m(\Omega; \mathbb{R}^d)$, that satisfy $\|f\|_{H^m(\Omega; \mathbb{R}^d)} < \kappa_1$ and $\|g - y_{A_0}\|_{H^{m+2}(\Omega; \mathbb{R}^d)} < \kappa_1$, the problem*

$$-\mathrm{div}(DW_{\mathrm{CB}}(\nabla y(x))) = f(x), \quad \text{if } x \in \Omega,$$
$$y(x) = g(x), \quad \text{if } x \in \partial\Omega,$$

*has exactly one weak solution with $\|y - g\|_{H^{m+2}(\Omega; \mathbb{R}^d)} < \kappa_2$. Furthermore, we always have*

$$\sup_x |((\nabla y(x) - A_0)\rho)_{\rho \in \mathcal{R}}| < r_0,$$

*that $y$ is a $W^{1,\infty}$-local minimizer of $E(\cdot; f; g)$ and that $y$ depends $C^1$ on $f$ and $g$ in the norms used above.*

*Proof of Theorem 4.4.* Let $X = H^{m+2}(\Omega; \mathbb{R}^d)$, $X_0 = H^{m+2}(\Omega; \mathbb{R}^d) \cap H_0^1(\Omega; \mathbb{R}^d)$ and $Y = H^m(\Omega; \mathbb{R}^d)$. Define $T \colon B_{r_1}^{X_0}(0) \times B_{r_2}^{X}(0) \times Y \to Y$,

$$T(u, h, f) = -\mathrm{div}(DW_{\mathrm{CB}}(A_0 + \nabla h(x) + \nabla u(x))) - f(x).$$

If we choose $r_1, r_2 > 0$ small enough, then we always have

$$\sup_{x \in \Omega} |((\nabla h(x) + \nabla u(x))\rho)_{\rho \in \mathcal{R}}| < r_0.$$

since $H^{m+2}(\Omega;\mathbb{R}^d) \hookrightarrow C_b^1(\Omega;\mathbb{R}^d)$. Using the properties of $F$ from Lemma B.5 with $V = B_{r_0}((A_0\rho)_{\rho\in\mathcal{R}})$, this implies that $T$ is well-defined, is in $C^1$ and

$$\partial_u T(u,h,f)[v](x) = -\operatorname{div}(D^2W_{\mathrm{CB}}(A_0 + \nabla u(x) + \nabla h(x))[\nabla v(x)]).$$

In particular,

$$\partial_u T(0,0,0)[v](x) = -\operatorname{div}(D^2W_{\mathrm{CB}}(A_0)[\nabla v(x)]).$$

Since $D^2W_{\mathrm{CB}}(A_0)$ satisfies the Legendre-Hadamard condition, the invertibility of $\partial_u T(0,0,0)\colon X_0 \to Y$ follows from Corollary 4.3. Now the main statement on existence, uniqueness and $C^1$-dependence follows from a standard Banach space implicit function theorem, as can be found, e.g., in [Dei10, Thm. 15.1 and Cor. 15.1], and then setting $g = y_{A_0} + h$, $y = y_{A_0} + h + u$.

Furthermore, if we choose $r_1, r_2$ even smaller, the above statements are still true and we can achieve that

$$\sup_{x\in\Omega}|\nabla h(x) + \nabla u(x)| < \tilde{r}$$

for all $(u,h) \in B_{r_1}^{X_0}(0) \times B_{r_2}^{X}(0)$, where $\tilde{r}$ is such that

$$\int_\Omega D^2W_{\mathrm{CB}}(\nabla z(x))[\nabla s, \nabla s]\,dx \geq \frac{\lambda_{\mathrm{LH}}(A_0)}{2}\int_\Omega |\nabla s(x)|^2\,dx$$

holds for all $s \in H_0^1(\Omega;\mathbb{R}^d)$ and $z \in W^{1,\infty}(\Omega;\mathbb{R}^d)$ with $|\nabla z(x) - A_0| \leq \tilde{r}$ a.e.. This is possible since $D^2W_{\mathrm{CB}}$ is uniformly continuous.

Now, if $w \in W^{1,\infty}(\Omega;\mathbb{R}^d)$ is in this space close enough to $y$, then

$$|\nabla w(x) - A_0| \leq \tilde{r},$$

and we have

$$E(w;f;g) = E(y;f;g)$$

$$+ \int_0^1\int_\Omega (1-t)D^2W_{\mathrm{CB}}((1-t)\nabla y(x) + t\nabla w(x))[\nabla w(x) - \nabla y(x)]^2\,dx\,dt$$

$$\geq E(y;f;g) + \frac{\lambda}{2}\int_\Omega |\nabla w(x) - \nabla y(x)|^2\,dx.$$

Hence, $y$ is a $W^{1,\infty}$-local minimizer of $E(\cdot;f;g)$ (strongly with respect to the $H_0^1(\Omega;\mathbb{R}^d)$-Norm). $\qquad\square$

## 4.2 Existence and Convergence of Solutions to the Atomistic Equations

Given $\varepsilon \in (0,1]$ and $f \in L^2(\Omega)$ we will write

$$\tilde{f}(x) = \fint_{Q_\varepsilon(x)} f(z)\,dz$$

for $x \in \mathrm{int}_\varepsilon\,\Omega$. If $\Omega$ has Lipschitz boundary and we have a deformation $y \in H^1(\Omega; \mathbb{R}^d)$ we will write

$$S_\varepsilon y(x) = \eta_\varepsilon * (y_{A_0} + E(y - y_{A_0}))(x)$$

for $x \in \varepsilon\mathbb{Z}^d$, where $\eta_\varepsilon$ is the standard scaled smoothing kernel and $E$ is an extension operator for all Sobolev spaces, see [Ste70, Chapter VI], such that every $Eu$ has support in a fixed ball $B_{R_E}(0)$. In the following theorem $\tilde{f}$ and $S_\varepsilon y$ are our reference points for the atomistic body forces, boundary conditions and deformations.

**Theorem 4.5.** *Let $d \in \{1, 2, 3, 4\}$ and let $\Omega \subset \mathbb{R}^d$ be an open, bounded set with $C^4$-boundary. Let $r_0 > 0$, $W_{\mathrm{atom}} \in C^5(\overline{B_{r_0}((A_0\rho)_{\rho \in \mathcal{R}})})$ and assume $\lambda_{\mathrm{atom}}(A_0) > 0$. Then there are constants $K_1, K_2, K_3 > 0$ with $K_1 < \kappa_1$ such that for every $f \in H^2(\Omega; \mathbb{R}^d)$ with $\|f\|_{H^2(\Omega;\mathbb{R}^d)} \leq K_1$, $g \in H^4(\Omega; \mathbb{R}^d)$ with $\|g - y_{A_0}\|_{H^4(\Omega;\mathbb{R}^d)} \leq K_1$, $\varepsilon \in (0, 1]$, $\gamma \in [\frac{d}{2}, 2]$, $f_{\mathrm{atom}} \colon \mathrm{int}_\varepsilon\,\Omega \to \mathbb{R}^d$ with $\|f_{\mathrm{atom}} - \tilde{f}\|_{h_\varepsilon^{-1}(\mathrm{int}_\varepsilon\,\Omega)} \leq K_2\varepsilon^\gamma$ and $g_{\mathrm{atom}} \colon \partial_\varepsilon\Omega \to \mathbb{R}^d$ with $\|g_{\mathrm{atom}} - S_\varepsilon y\|_{\partial_\varepsilon\Omega} \leq K_2\varepsilon^\gamma$, where $y$ is given by Theorem 4.4, there is a unique $y_{\mathrm{atom}} \in \mathcal{A}_\varepsilon(\Omega, g_{\mathrm{atom}})$ with $\|y_{\mathrm{atom}} - S_\varepsilon y\|_{h_\varepsilon^1(\mathrm{sint}_\varepsilon\,\Omega)} \leq K_3\varepsilon^\gamma$ such that*

$$-\operatorname{div}_{\mathcal{R},\varepsilon}\big(DW_{\mathrm{atom}}(D_{\mathcal{R},\varepsilon}y_{\mathrm{atom}}(x))\big) = f_{\mathrm{atom}}(x)$$

*for all $x \in \mathrm{int}_\varepsilon\,\Omega$. Furthermore, $y_{\mathrm{atom}}$ is a strict local minimizer of $E_\varepsilon(\cdot, f_{\mathrm{atom}}, g_{\mathrm{atom}})$.*

   *Additionally, there is a $K_4 > 0$ such that whenever $\gamma \in (1, 2]$ and $E(y - y_{A_0}) \in C^{2,(\gamma-1)}(\Omega)$ then*

$$\|y_{\mathrm{atom}} - y\|_{h_\varepsilon^1(\mathrm{int}_\varepsilon\,\Omega)} \leq (K_3 + K_4|\nabla^2 E(y - y_{A_0})|_{\gamma-1})\varepsilon^\gamma.$$

*Remark 4.6.* If $d = 3$ and $\gamma = \frac{3}{2}$ the assumption in the additional statement is automatically satisfied since $E(y - y_{A_0}) \in H^4(B_{R_E}(0)) \hookrightarrow C^{2,\frac{1}{2}}(B_{R_E}(0))$.

The proof relies on the following quantitative implicit function theorem:

**Theorem 4.7.** *Let $X$ be a Banach space and $Y, Z$ normed spaces, $U \subset X$, $V \subset Y$ open and $F \colon U \times V \to Z$ Fréchet-differentiable. Furthermore, assume that there are $\rho, \tau, \kappa_1, \kappa_2, \kappa_3 > 0$ and functions $\omega_1, \omega_2 \colon [0, \infty)^2 \to [0, \infty)$, non-decreasing in both variables, such that $\overline{B_\rho(0)} \subset U$, $\overline{B_\tau(0)} \subset V$,*

$$\|F(0,0)\|_Z \le \kappa_1,$$
$$\|\partial_u F(0,0)^{-1}\|_{L(Z,X)} \le \kappa_2,$$
$$\|\partial_h F(0,0)\|_{L(Y,Z)} \le \kappa_3,$$
$$\|\partial_u F(0,0) - \partial_u F(u,h)\|_{L(X,Z)} \le \omega_1(\|u\|_X, \|h\|_Y)$$
$$\forall (u,h) \in \overline{B_\rho(0)} \times \overline{B_\tau(0)},$$
$$\|\partial_h F(0,0) - \partial_h F(u,h)\|_{L(Y,Z)} \le \omega_2(\|u\|_X, \|h\|_Y)$$
$$\forall (u,h) \in \overline{B_\rho(0)} \times \overline{B_\tau(0)},$$

$$2\kappa_2 \omega_1(\rho, \tau) \le 1$$

*and*

$$2\kappa_2 \kappa_1 + 2\tau \kappa_2 \kappa_3 + 2\tau \kappa_2 \omega_2(\rho, \tau) \le \rho.$$

*Then, for every $h \in \overline{B_\tau(0)}$ there is a unique $u \in \overline{B_\rho(0)}$ with $F(u,h) = 0$.*

*Proof.* Let

$$G_h(u) = u - \partial_u F(0,0)^{-1} F(u,h).$$

If $\|u\| \le \rho$ and $\|h\| \le \tau$ then

$$G_h(u) = \partial_u F(0,0)^{-1} \big( - F(0,0) - \partial_h F(0,0) h$$
$$+ F(0,0) + DF(0,0)[u,h] - F(u,h) \big).$$

Therefore,

$$\|G_h(u)\| \le \kappa_2(\kappa_1 + \kappa_3 \tau) + \kappa_2 \left\| \int_0^1 (DF(0,0) - DF(tu, th))[u,h]\, dt \right\|$$
$$\le \kappa_2(\kappa_1 + \kappa_3 \tau + \omega_1(\rho, \tau)\rho + \omega_2(\rho, \tau)\tau) \le \rho.$$

Furthermore, for $u, v \in \overline{B_\rho(0)}$ we have

$$\|G_h(u) - G_h(v)\|$$
$$\leq \kappa_2 \left\| \int_0^1 (\partial_u F(u + t(v - u), h) - \partial_u F(0,0))(v - u)\, dt \right\|$$
$$\leq \kappa_2 \omega_1(\rho, \tau) \|v - u\|$$
$$\leq \frac{1}{2} \|v - u\|.$$

Hence, $G_h$ has a unique fixed point in $u \in \overline{B_\rho(0)}$.                  □

More precisely, we want to use the following corollary:

**Corollary 4.8.** *Let $d \in \{1, 2, 3, 4\}$. Assume we have a family $F_\varepsilon \colon U_\varepsilon \times V_\varepsilon \to Z_\varepsilon$ with $\varepsilon \in (0, 1]$ and $U_\varepsilon \subset X_\varepsilon, V_\varepsilon \subset Y_\varepsilon$ open, where $Y_\varepsilon, Z_\varepsilon$ are normed spaces and $X_\varepsilon$ Banach spaces. Furthermore, assume that the $F_\varepsilon$ are Fréchet-differentiable and we have fixed $r_1, r_2 > 0$ such that $\overline{B_{r_1 \varepsilon^{\frac{d}{2}}}(0)} \subset U_\varepsilon$ and $\overline{B_{r_2 \varepsilon^{\frac{d}{2}}}(0)} \subset V_\varepsilon$. Now, assume there are constants $A, M_1, M_2, M_3, M_4 > 0$, such that*

$$\|F_\varepsilon(0,0)\|_{Z_\varepsilon} \leq A\varepsilon^2,$$
$$\|\partial_u F_\varepsilon(0,0)^{-1}\|_{L(Z_\varepsilon, X_\varepsilon)} \leq M_1,$$
$$\|\partial_h F_\varepsilon(0,0)\|_{L(Y_\varepsilon, Z_\varepsilon)} \leq M_2,$$
$$\|\partial_u F_\varepsilon(0,0) - \partial_u F_\varepsilon(u,h)\|_{L(X_\varepsilon, Z_\varepsilon)} \leq M_3 \varepsilon^{-\frac{d}{2}}(\|u\|_{X_\varepsilon} + \|h\|_{Y_\varepsilon})$$
$$\forall (u,h) \in \overline{B_{r_1 \varepsilon^{\frac{d}{2}}}(0)} \times \overline{B_{r_2 \varepsilon^{\frac{d}{2}}}(0)},$$
$$\|\partial_h F_\varepsilon(0,0) - \partial_h F_\varepsilon(u,h)\|_{L(Y_\varepsilon, Z_\varepsilon)} \leq M_4 \varepsilon^{-\frac{d}{2}}(\|u\|_{X_\varepsilon} + \|h\|_{Y_\varepsilon})$$
$$\forall (u,h) \in \overline{B_{r_1 \varepsilon^{\frac{d}{2}}}(0)} \times \overline{B_{r_2 \varepsilon^{\frac{d}{2}}}(0)},$$

*and*

$$A \leq \min \left\{ \frac{r_1}{4M_1}, \frac{1}{16 M_1^2 M_3} \right\}.$$

*If we now set $\rho_\varepsilon = \lambda_1 \varepsilon^\gamma$ and $\tau_\varepsilon = \lambda_2 \varepsilon^\gamma$ for arbitrary $\gamma \in \left[ \frac{d}{2}, 2 \right]$ and*

$$\lambda_1 = \min \left\{ r_1, \frac{1}{4 M_1 M_3} \right\},$$
$$\lambda_2 = \min \left\{ r_2, \frac{1}{4 M_1 M_3}, \frac{\lambda_1 M_3}{4 M_1 M_2 M_3 + 2 M_4} \right\},$$

*then for every $\varepsilon \in (0, 1]$ and every $h_\varepsilon \in \overline{B_{\tau_\varepsilon}(0)}$ there is a unique $u_\varepsilon \in \overline{B_{\rho_\varepsilon}(0)}$ with $F_\varepsilon(u_\varepsilon, h_\varepsilon) = 0$.*

*Proof.* We set

$$\kappa_1 = A\varepsilon^2,$$
$$\kappa_2 = M_1,$$
$$\kappa_3 = M_2,$$
$$\omega_1(s,t) = M_3\varepsilon^{-\frac{d}{2}}(s+t),$$
$$\omega_2(s,t) = M_4\varepsilon^{-\frac{d}{2}}(s+t),$$

and calculate

$$2\kappa_2\omega_1(\rho_\varepsilon,\lambda_\varepsilon) = 2M_1M_3\varepsilon^{\gamma-\frac{d}{2}}(\lambda_1+\lambda_2) \leq 1$$

and

$$2\kappa_2\kappa_1 + 2\tau\kappa_2\kappa_3 + 2\tau\kappa_2\omega_2(\rho_\varepsilon,\tau_\varepsilon)$$
$$= 2M_1A\varepsilon^2 + 2M_1\lambda_2\varepsilon^\gamma(M_2 + M_4\varepsilon^{\gamma-\frac{d}{2}}(\lambda_1+\lambda_2))$$
$$\leq \frac{\lambda_1}{2}\varepsilon^\gamma + \frac{2\lambda_1M_1M_3}{4M_1M_2M_3+2M_4}\varepsilon^\gamma\left(M_2 + \frac{M_4}{2M_1M_3}\right)$$
$$= \rho_\varepsilon$$

$\square$

Thus, we can find a solution to the discrete equations by finding an approximate solution with good estimates on the residuum, invertibility and continuity of the derivatives. The approximate solution in our case will be a smooth approximation of the solution to the corresponding continuous equations with the Cauchy-Born energy density.

## 4.3   Residual Estimates

As discussed before, one of the key ingredient to our results are the residual estimates. Formally, one applies the atomistic equations to a smooth solution $y$ of the Cauchy-Born continuum equations and then expands the solution as well as the nonlinearity in $\varepsilon$ with Taylor's theorem. One then finds that all the terms of negative order in $\varepsilon$ cancel. The $\varepsilon^0$-terms vanish because they correspond precisely to the continuum equations. And all terms on order $\varepsilon^1$ cancel under mild symmetry assumptions (on a Bravais lattice). The residuum of $y$ should therefore be on the order $\varepsilon^2$. It is not too difficult to make these arguments rigorous if $y$ has quite

high regularity. To drastically reduce these regularity assumptions, we
will prove finer estimates that focus on averaging and regularizing on
the atomistic scale. In the end, we will indeed discuss the residuum of
a regularization of the continuum solution and not the residuum at the
continuum solution itself.

**Proposition 4.9.** *Let* $V \subset \mathbb{R}^{d\times\mathcal{R}}$ *be open and* $W_{\mathrm{atom}} \in C_b^4(V)$. *Let* $f \in L^2(\Omega;\mathbb{R}^d)$ *and set as before*

$$\tilde{f}(x) = \fint_{Q_\varepsilon(x)} f(a)\, da$$

*for* $x \in \mathrm{int}_\varepsilon\, \Omega$. *Furthermore let* $\varepsilon \in (0,1]$ *and* $y \in C^{3,1}(\mathbb{R}^d;\mathbb{R}^d)$ *with*

$$\mathrm{co}\{D_{\mathcal{R},\varepsilon}y(\hat{x}+\varepsilon\sigma),\, (\nabla y(x)\rho)_{\rho\in\mathcal{R}}\} \subset V$$

*for all* $x \in \Omega_\varepsilon$ *and* $\sigma \in \mathcal{R}\cup\{0\}$. *Then we have*

$$\left\| -\tilde{f} - \mathrm{div}_{\mathcal{R},\varepsilon}\left(DW_{\mathrm{atom}}(D_{\mathcal{R},\varepsilon}y)\right) \right\|_{\ell_\varepsilon^2(\mathrm{int}_\varepsilon\,\Omega)}$$
$$\leq \left\| -f - \mathrm{div}\, DW_{\mathrm{CB}}(\nabla y)\right\|_{L^2(\Omega_\varepsilon;\mathbb{R}^d)} + C\varepsilon^2 \Big\| \|\nabla^4 y\|_{L^\infty(B_{\varepsilon R}(x))}$$
$$+ \|\nabla^3 y\|_{L^\infty(B_{\varepsilon R}(x))}^{\frac{3}{2}} + \|\nabla^2 y\|_{L^\infty(B_{\varepsilon R}(x))}^3 + \varepsilon\|\nabla^3 y\|_{L^\infty(B_{\varepsilon R}(x))}^2 \Big\|_{L^2(\Omega_\varepsilon)},$$

*where* $\Omega_\varepsilon = \bigcup_{z\in\mathrm{int}_\varepsilon\,\Omega} Q_\varepsilon(z)$, $R = 2R_{\max} + \frac{3\sqrt{d}}{2}$ *and* $C > 0$ *with* $C = C(d,\mathcal{R}, \|D^2 W_{\mathrm{atom}}\|_{C^2(V)})$.

*Proof.* For $x \in \Omega_\varepsilon$, $\sigma \in \mathcal{R}$ and $\rho \in \mathcal{R}\cup\{0\}$ set

$$r_{1,\varepsilon}(x;\sigma,\rho) = \frac{y(\hat{x}-\varepsilon\rho+\varepsilon\sigma) - y(\hat{x}-\varepsilon\rho)}{\varepsilon} - \nabla y(x)\sigma$$
$$r_{2,\varepsilon}(x;\sigma,\rho) = \frac{y(\hat{x}-\varepsilon\rho+\varepsilon\sigma) - y(\hat{x}-\varepsilon\rho)}{\varepsilon} - \nabla y(x)\sigma$$
$$- \frac{1}{2}\varepsilon\nabla^2 y(x)[\sigma-\rho+\frac{\hat{x}-x}{\varepsilon}, \sigma-\rho+\frac{\hat{x}-x}{\varepsilon}]$$
$$+ \frac{1}{2}\varepsilon\nabla^2 y(x)[-\rho+\frac{\hat{x}-x}{\varepsilon}, -\rho+\frac{\hat{x}-x}{\varepsilon}]$$

$$r_{3,\varepsilon}(x;\sigma,\rho) = \frac{y(\hat{x}-\varepsilon\rho+\varepsilon\sigma)-y(\hat{x}-\varepsilon\rho)}{\varepsilon} - \nabla y(x)\sigma$$

$$-\frac{1}{2}\varepsilon\nabla^2 y(x)[\sigma-\rho+\frac{\hat{x}-x}{\varepsilon},\sigma-\rho+\frac{\hat{x}-x}{\varepsilon}]$$

$$+\frac{1}{2}\varepsilon\nabla^2 y(x)[-\rho+\frac{\hat{x}-x}{\varepsilon},-\rho+\frac{\hat{x}-x}{\varepsilon}]$$

$$-\frac{1}{6}\varepsilon^2\nabla^3 y(x)[\sigma-\rho+\frac{\hat{x}-x}{\varepsilon},\sigma-\rho+\frac{\hat{x}-x}{\varepsilon},\sigma-\rho+\frac{\hat{x}-x}{\varepsilon}]$$

$$+\frac{1}{6}\varepsilon^2\nabla^3 y(x)[-\rho+\frac{\hat{x}-x}{\varepsilon},-\rho+\frac{\hat{x}-x}{\varepsilon},-\rho+\frac{\hat{x}-x}{\varepsilon}].$$

First order Taylor expansions with integral remainder of $y(\hat{x}-\varepsilon\rho+\varepsilon\sigma)$ and $y(\hat{x}-\varepsilon\rho)$ at $x$ give the estimate

$$|r_{1,\varepsilon}(x;\sigma,\rho)| \le \varepsilon\bar{R}^2\|\nabla^2 y\|_{L^\infty(B_{\varepsilon\bar{R}}(x))},$$

where $\bar{R} = 2R_{max} + \frac{1}{2}\sqrt{d}$ and we use that $|\hat{x}-x| \le \frac{1}{2}\varepsilon\sqrt{d}$. Similarly, second and third order Taylor expansions give

$$|r_{2,\varepsilon}(x;\sigma,\rho)| \le \frac{1}{3}\varepsilon^2\bar{R}^3\|\nabla^3 y\|_{L^\infty(B_{\varepsilon\bar{R}}(x))}$$

$$|r_{3,\varepsilon}(x;\sigma,\rho)| \le \frac{1}{12}\varepsilon^3\bar{R}^4\|\nabla^4 y\|_{L^\infty(B_{\varepsilon\bar{R}}(x))}.$$

Now, doing a second order Taylor expansion with integral remainder of $DW_{\text{atom}}$ at $(\nabla y(x)\rho)_{\rho\in\mathcal{R}}$, using the definition of $r_{3,\varepsilon}$ in the first order term, the definition of $r_{2,\varepsilon}$ in the second order term and the definition of $r_{1,\varepsilon}$ in the remainder and then collecting the terms with the same exponent in $\varepsilon$ gives

$$-f(x) - \text{div}_{\mathcal{R},\varepsilon}\left(DW_{\text{atom}}(D_{\mathcal{R},\varepsilon}y(\hat{x}))\right) = \varepsilon^{-1}I_{-1} + \varepsilon^0 I_0 + \varepsilon^1 I_1 + R_\varepsilon(x),$$

where

$$I_{-1} = -\sum_{\rho\in\mathcal{R}} D_{e_\rho}W_{\text{atom}}((\nabla y(x)\rho)_{\rho\in\mathcal{R}}) - D_{e_\rho}W_{\text{atom}}((\nabla y(x)\rho)_{\rho\in\mathcal{R}}) = 0$$

and

$$(I_0)_j = -f_j(x) - \sum_{\rho,\sigma\in\mathcal{R}} D^2 W_{\text{atom}}((\nabla y(x)\rho)_{\rho\in\mathcal{R}})\Big[e_j\otimes e_\rho,$$

$$\frac{1}{2}\nabla^2 y(x)\Big([\sigma+\frac{\hat{x}-x}{\varepsilon}]^2 - [\frac{\hat{x}-x}{\varepsilon}]^2$$

$$- [\sigma-\rho+\frac{\hat{x}-x}{\varepsilon}]^2 + [-\rho+\frac{\hat{x}-x}{\varepsilon}]^2\Big)\otimes e_\sigma\Big]$$

$$= -f_j(x) - \sum_{\rho,\sigma\in\mathcal{R}} D^2 W_{\text{atom}}((\nabla y(x)\rho)_{\rho\in\mathcal{R}}) \Big[ e_j \otimes e_\rho,$$

$$\nabla^2 y(x)[\sigma,\rho] \otimes e_\sigma \Big]$$

$$= -f_j(x) - \sum_i \sum_{\rho,\sigma\in\mathcal{R}} D^2 W_{\text{atom}}((\nabla y(x)\rho)_{\rho\in\mathcal{R}}) \Big[ ((e_j \otimes e_i)\rho) \otimes e_\rho,$$

$$\nabla^2 y(x)[\sigma,e_i] \otimes e_\sigma \Big]$$

$$= -f_j(x) - \sum_i \frac{\partial}{\partial x_i} \sum_{\rho\in\mathcal{R}} DW_{\text{atom}}((\nabla y(x)\rho)_{\rho\in\mathcal{R}}) \Big[ ((e_j \otimes e_i)\rho) \otimes e_\rho \Big]$$

$$= -f_j(x) - \sum_i \frac{\partial}{\partial x_i} DW_{\text{CB}}(\nabla y(x))[e_j \otimes e_i]$$

$$= \big( -f(x) - \operatorname{div} DW_{\text{CB}}(\nabla y(x)) \big)_j$$

and

$$(I_1)_j = - \sum_{\rho,\sigma\in\mathcal{R}} D^2 W_{\text{atom}}((\nabla y(x)\rho)_{\rho\in\mathcal{R}}) \Big[ e_j \otimes e_\rho,$$

$$\frac{1}{6}\nabla^3 y(x)\Big( [\sigma + \frac{\hat{x}-x}{\varepsilon}]^3 - [\frac{\hat{x}-x}{\varepsilon}]^3$$

$$- [\sigma - \rho + \frac{\hat{x}-x}{\varepsilon}]^3 + [-\rho + \frac{\hat{x}-x}{\varepsilon}]^3 \Big) \otimes e_\sigma \Big]$$

$$- \frac{1}{2} \sum_{\rho,\sigma,\tau\in\mathcal{R}} D^3 W_{\text{atom}}((\nabla y(x)\rho)_{\rho\in\mathcal{R}}) \big[ e_j \otimes e_\rho \big]$$

$$\Big( \Big[ \frac{1}{2}\nabla^2 y(x)\Big( [\sigma + \frac{\hat{x}-x}{\varepsilon}]^2 - [\frac{\hat{x}-x}{\varepsilon}]^2 \Big) \otimes e_\sigma,$$

$$\frac{1}{2}\nabla^2 y(x)\Big( [\tau + \frac{\hat{x}-x}{\varepsilon}]^2 - [\frac{\hat{x}-x}{\varepsilon}]^2 \Big) \otimes e_\tau \Big]$$

$$- \Big[ \frac{1}{2}\nabla^2 y(x)\Big( [\sigma - \rho + \frac{\hat{x}-x}{\varepsilon}]^2 - [\frac{\hat{x}-x}{\varepsilon} - \rho]^2 \Big) \otimes e_\sigma,$$

$$\frac{1}{2}\nabla^2 y(x)\Big( [\tau - \rho + \frac{\hat{x}-x}{\varepsilon}]^2 - [\frac{\hat{x}-x}{\varepsilon} - \rho]^2 \Big) \otimes e_\tau \Big] \Big)$$

$$= - \sum_{\rho,\sigma\in\mathcal{R}} D^2 W_{\text{atom}}((\nabla y(x)\rho)_{\rho\in\mathcal{R}}) \Big[ e_j \otimes e_\rho,$$

$$\frac{1}{2}\nabla^3 y(x)[\sigma,\rho,\sigma - \rho + 2\frac{\hat{x}-x}{\varepsilon}] \otimes e_\sigma \Big]$$

$$-\frac{1}{2}\sum_{\rho,\sigma,\tau\in\mathcal{R}}D^3W_{\text{atom}}((\nabla y(x)\rho)_{\rho\in\mathcal{R}})\big[e_j\otimes e_\rho\big]$$

$$\Bigg(\Big[\nabla^2y(x)[\sigma,\rho]\otimes e_\sigma,\nabla^2y(x)[\tau,\tfrac{1}{2}\tau+\tfrac{\hat{x}-x}{\varepsilon}]\otimes e_\tau\Big]$$

$$+\Big[\nabla^2y(x)[\sigma,\tfrac{1}{2}\sigma+\tfrac{\hat{x}-x}{\varepsilon}]\otimes e_\sigma,\nabla^2y(x)[\tau,\rho]\otimes e_\tau\Big]$$

$$-\Big[\nabla^2y(x)[\sigma,\rho]\otimes e_\sigma,\nabla^2y(x)[\tau,\rho]\otimes e_\tau\Big]\Bigg)$$

$$=-\sum_{\rho,\sigma\in\mathcal{R}}D^2W_{\text{atom}}((\nabla y(x)\rho)_{\rho\in\mathcal{R}})\Big[e_j\otimes e_\rho,$$

$$\nabla^3y(x)[\sigma,\rho,\tfrac{\hat{x}-x}{\varepsilon}]\otimes e_\sigma\Big]$$

$$-\frac{1}{2}\sum_{\rho,\sigma,\tau\in\mathcal{R}}D^3W_{\text{atom}}((\nabla y(x)\rho)_{\rho\in\mathcal{R}})\big[e_j\otimes e_\rho\big]$$

$$\Bigg(\Big[\nabla^2y(x)[\sigma,\rho]\otimes e_\sigma,\nabla^2y(x)[\tau,\tfrac{\hat{x}-x}{\varepsilon}]\otimes e_\tau\Big]$$

$$+\Big[\nabla^2y(x)[\sigma,\tfrac{\hat{x}-x}{\varepsilon}]\otimes e_\sigma,\nabla^2y(x)[\tau,\rho]\otimes e_\tau\Big]\Bigg)$$

where in the last equality we applied the symmetry condition in the form of Lemma 2.1. While the last expression is not necessarily zero, it is linear in $\frac{\hat{x}-x}{\varepsilon}$, with coefficients depending on $x$. Therefore, the average $\frac{1}{2}(I_1(x)+I_1(\bar{x}))$ is actually of higher order. Here $\bar{x}$ denotes the almost everywhere uniquely defined point in the same cube as $x$ such that $\frac{1}{2}(x+\bar{x})=\hat{x}$. To be more precise, we have

$$\left|\frac{\varepsilon}{2}(I_1(x)+I_1(\bar{x}))\right|$$

$$\leq\frac{1}{2}\varepsilon^2|\mathcal{R}|^2\|D^2W_{\text{atom}}\|_\infty\sqrt{d}\|\nabla^4y\|_{L^\infty(B_{\varepsilon\sqrt{d}}(x))}R_{\max}^2\frac{\sqrt{d}}{2}$$

$$+\frac{1}{2}\varepsilon^2|\mathcal{R}|^{\frac{5}{2}}\|D^3W_{\text{atom}}\|_\infty\sqrt{d}\|\nabla^2y\|_{L^\infty(B_{\varepsilon\sqrt{d}}(x))}R_{\max}^3\frac{\sqrt{d}}{2}|\nabla^3y(x)|$$

$$+\varepsilon^2|\mathcal{R}|^3\|D^3W_{\text{atom}}\|_\infty\sqrt{d}\|\nabla^3y\|_{L^\infty(B_{\varepsilon\sqrt{d}}(x))}\|\nabla^2y\|_{L^\infty(B_{\varepsilon\sqrt{d}}(x))}R_{\max}^3\frac{\sqrt{d}}{2}$$

$$+\varepsilon^2|\mathcal{R}|^{\frac{7}{2}}\|D^4W_{\text{atom}}\|_\infty\sqrt{d}\|\nabla^2y\|_{L^\infty(B_{\varepsilon\sqrt{d}}(x))}R_{\max}^4\frac{\sqrt{d}}{2}|\nabla^2y(x)|^2$$

$$\leq C\varepsilon^2\Big(\|\nabla^4y\|_{L^\infty(B_{\varepsilon\sqrt{d}}(x))}+\|\nabla^3y\|_{L^\infty(B_{\varepsilon\sqrt{d}}(x))}^{\frac{3}{2}}+\|\nabla^2y\|_{L^\infty(B_{\varepsilon\sqrt{d}}(x))}^3\Big).$$

Using the bounds we have on $r_{1,\varepsilon}$, $r_{2,\varepsilon}$ and $r_{3,\varepsilon}$, we estimate

$$|(R_\varepsilon)_j(x)|$$
$$\leq \varepsilon^2 \Big( |\mathcal{R}|^2 \|D^2 W_{\text{atom}}\|_\infty \bar{R}^4 \frac{1}{6} \|\nabla^4 y\|_{L^\infty(B_{\varepsilon \bar{R}}(x))}$$
$$+ |\mathcal{R}|^3 \|D^3 W_{\text{atom}}\|_\infty \frac{2}{3} \bar{R}^3 R_{\max}(R_{\max} + \frac{\sqrt{d}}{2}) |\nabla^2 y(x)| \|\nabla^3 y\|_{L^\infty(B_{\varepsilon \bar{R}}(x))}$$
$$+ |\mathcal{R}|^4 \|D^4 W_{\text{atom}}\|_\infty \frac{1}{3} \bar{R}^6 \|\nabla^2 y\|_{L^\infty(B_{\varepsilon \bar{R}}(x))}^3 \Big)$$
$$+ \varepsilon^3 \Big( |\mathcal{R}|^3 \|D^3 W_{\text{atom}}\|_\infty \bar{R}^6 \frac{1}{9} \|\nabla^3 y\|_{L^\infty(B_{\varepsilon \bar{R}}(x))}^2 \Big)$$
$$\leq C \varepsilon^2 \big( \|\nabla^4 y\|_{L^\infty(B_{\varepsilon \bar{R}}(x))} + \|\nabla^3 y\|_{L^\infty(B_{\varepsilon \bar{R}}(x))}^{\frac{3}{2}}$$
$$+ \|\nabla^2 y\|_{L^\infty(B_{\varepsilon \bar{R}}(x))}^3 + \varepsilon \|\nabla^3 y\|_{L^\infty(B_{\varepsilon \bar{R}}(x))}^2 \big)$$

Combining these estimates and using $\bar{R} + \sqrt{d} = R$, we get

$$\Big| -\frac{f(x) + f(\bar{x})}{2} - \text{div}_{\mathcal{R},\varepsilon} \big( DW_{\text{atom}}(D_{\mathcal{R},\varepsilon} y(\hat{x})) \big) \Big|$$
$$\leq \Big| \frac{-f(x) - \text{div}\, DW_{\text{CB}}(\nabla y(x)) - f(\bar{x}) - \text{div}\, DW_{\text{CB}}(\nabla y(\bar{x}))}{2} \Big|$$
$$+ C \varepsilon^2 \Big( \|\nabla^4 y\|_{L^\infty(B_{\varepsilon R}(x))} + \|\nabla^3 y\|_{L^\infty(B_{\varepsilon R}(x))}^{\frac{3}{2}}$$
$$+ \|\nabla^2 y\|_{L^\infty(B_{\varepsilon R}(x))}^3 + \varepsilon \|\nabla^3 y\|_{L^\infty(B_{\varepsilon R}(x))}^2 \Big).$$

But,

$$\varepsilon^d \sum_{z \in \text{int}_\varepsilon \Omega} \big( -\tilde{f}(z) - \text{div}_{\mathcal{R},\varepsilon} \big( DW_{\text{atom}}(D_{\mathcal{R},\varepsilon} y(z)) \big) \big)^2$$
$$\leq \sum_{z \in \text{int}_\varepsilon \Omega} \int_{Q_\varepsilon(z)} \Big( -\frac{f(a) + f(\bar{a})}{2} - \text{div}_{\mathcal{R},\varepsilon} \big( DW_{\text{atom}}(D_{\mathcal{R},\varepsilon} y(\hat{a})) \big) \Big)^2 da,$$

which combined gives the desired result.                                    $\square$

These residual estimates are particularly strong if we combine them with the following two approximation results:

**Proposition 4.10.** *For any $R > 0$, $k, d \in \mathbb{N}$, $p \geq 1$, there is a $C = C(R, d, p) > 0$ such that for any $U \subset \mathbb{R}^d$ measurable and $y \in W^{k,p}(U + B_{(R+1)\varepsilon}(0); \mathbb{R}^d)$ we have*

$$\Big\| \|\nabla^k(y * \eta_\varepsilon)\|_{L^\infty(B_{\varepsilon R}(\cdot))} \Big\|_{L^p(U)} \leq C \|\nabla^k y\|_{L^p(U + B_{(R+1)\varepsilon}(0))},$$

where $\eta_\varepsilon$ is the standard scaled smoothing kernel.

*Proof.* We directly calculate

$$\left\|\|\nabla^k(y * \eta_\varepsilon)\|_{L^\infty(B_{\varepsilon R}(\cdot))}\right\|_{L^p(U)}^p \leq \int_U \operatorname*{ess\,sup}_{z \in B_{\varepsilon R}(x)} \int_{\mathbb{R}^d} \eta_\varepsilon(a)|\nabla^k y(z + a)|^p \, da \, dx$$

$$\leq \|\eta\|_\infty \varepsilon^{-d} \int_U \int_{B_{\varepsilon(R+1)}(x)} |\nabla^k y(a)|^p \, da \, dx$$

$$\leq C(d)(R + 1)^d \int_{U + B_{\varepsilon(R+1)}(x)} |\nabla^k y(x)|^p \, dx.$$

$\square$

**Proposition 4.11.** *Let $d \in \{1, 2, 3, 4\}$, $\Omega \subset \mathbb{R}^d$ open and bounded with Lipschitz boundary, $V \subset \mathbb{R}^{d \times \mathcal{R}}$ be open and $W_{\mathrm{atom}} \in C_b^5(V)$. Then, there is a $C > 0$ such that for all $\varepsilon \in (0, 1]$ and all $y \in H^4(\Omega + B_\varepsilon(0); \mathbb{R}^d)$ with*

$$\inf_{x \in \Omega} \inf_{t \in [0,1]} \operatorname{dist}((1 - t)(\nabla y(x)\rho)_{\rho \in \mathcal{R}} + t(\nabla(y * \eta_\varepsilon)(x)\rho)_{\rho \in \mathcal{R}}, V^c) > 0,$$

*we have*

$$\|\operatorname{div} DW_{\mathrm{CB}}(\nabla y(x)) - \operatorname{div} DW_{\mathrm{CB}}(\nabla(y * \eta_\varepsilon)(x))\|_{L^2(\Omega)}$$
$$\leq C\varepsilon^2 \big(\|\nabla^2 y\|_{L^4(\Omega + B_\varepsilon(0))}\|\nabla^3 y\|_{L^4(\Omega + B_\varepsilon(0))} + \|\nabla^4 y\|_{L^2(\Omega + B_\varepsilon(0))}\big)$$

*where $\eta_\varepsilon$ is the standard scaled smoothing kernel.*

*Proof.* Now, since $\eta(z) = \eta(-z)$, we have

$$\nabla^k y(x) - \nabla^k(y * \eta_\varepsilon)(x)$$
$$= \int_{\mathbb{R}^d} \eta_\varepsilon(z)(\nabla^k y(x) - \nabla^k y(x + z)) \, dz$$
$$= \int_{\mathbb{R}^d} \eta_\varepsilon(z)(\nabla^k y(x) + \nabla^{k+1} y(x)[z] - \nabla^k y(x + z)) \, dz$$
$$= -\int_{\mathbb{R}^d} \int_0^1 \eta_\varepsilon(z)(1 - t)\nabla^{k+2} y(x + tz)[z, z] \, dt \, dz$$

But then

$$\int_\Omega |\nabla^k y(x) - \nabla^k(y * \eta_\varepsilon)(x)|^p \, dx$$

$$= \int_\Omega \Big| \int_{\mathbb{R}^d} \int_0^1 \eta_\varepsilon(z)(1-t)\nabla^{k+2}y(x+tz)[z,z] \, dt \, dz \Big|^p \, dx$$

$$\leq \varepsilon^{2p} \int_\Omega \int_{\mathbb{R}^d} \eta_\varepsilon(z) \int_0^1 |\nabla^{k+2}y(x+tz)|^p \, dt \, dz \, dx$$

$$\leq \varepsilon^{2p} \|\nabla^{k+2}y\|_{L^p(\Omega+B_\varepsilon(0))}^p.$$

While we used strong differentiability in the proof, the inequality extends directly to $W^{k+2,p}(\Omega + B_\varepsilon(0))$ by density. As in the proof of Lemma B.5 we get

$$DW_{\mathrm{CB}}(\nabla y) \in H^3(\Omega; \mathbb{R}^d),$$
$$DW_{\mathrm{CB}}(\nabla(y * \eta_\varepsilon)) \in H^3(\Omega; \mathbb{R}^d),$$
$$D^2 W_{\mathrm{CB}}((1-t)\nabla y + t\nabla(y * \eta_\varepsilon)) \in H^3(\Omega; \mathbb{R}^{d \times d}),$$

and thus

$$\|\operatorname{div} DW_{\mathrm{CB}}(\nabla y(x)) - \operatorname{div} DW_{\mathrm{CB}}(\nabla(y * \eta_\varepsilon)(x))\|_{L^2(\Omega)}$$

$$\leq \int_0^1 \|\operatorname{div} D^2 W_{\mathrm{CB}}((1-t)\nabla y + t\nabla(y * \eta_\varepsilon))[\nabla y - \nabla(y * \eta_\varepsilon)]\|_{L^2(\Omega)} \, dt$$

$$\leq C\big(\|\nabla^2 y - \nabla^2(y * \eta_\varepsilon)\|_{L^2(\Omega)}$$
$$+ \|\nabla y - \nabla(y * \eta_\varepsilon)\|_{L^4(\Omega)}(\|\nabla^2 y\|_{L^4(\Omega)} + \|\nabla^2(y * \eta_\varepsilon)\|_{L^4(\Omega)})\big)$$

$$\leq C\varepsilon^2 \big(\|\nabla^4 y\|_{L^2(\Omega+B_\varepsilon(0))} + \|\nabla^3 y\|_{L^4(\Omega+B_\varepsilon(0))}\|\nabla^2 y\|_{L^4(\Omega+B_\varepsilon(0))}\big),$$

where we used the inequality from above with $k = p = 2$ or $k = 1$ and $p = 4$, respectively. $\qquad\square$

## 4.4   Proof of the Main Result in the Static Case

We will need a discrete Poincaré-inequality:

**Proposition 4.12.** *Let $\Omega \subset \mathbb{R}^d$ be open and bounded. Then there is a $C_{\mathrm{P}}(\Omega) > 0$ such that for all $\varepsilon \in (0,1]$ and $u \in \mathcal{A}_\varepsilon(\Omega, 0)$ we have*

$$\|u\|_{\ell^2_\varepsilon(\mathrm{int}_\varepsilon \Omega)} \leq C_{\mathrm{P}}\|u\|_{h^1_\varepsilon(\mathrm{sint}_\varepsilon \Omega)}$$

*and for u:* $\mathrm{int}_\varepsilon \Omega \to \mathbb{R}^d$ *we have*

$$\|u\|_{h_\varepsilon^{-1}(\mathrm{int}_\varepsilon \Omega)} \leq C_\mathrm{P}\|u\|_{\ell_\varepsilon^2(\mathrm{int}_\varepsilon \Omega)}$$

*Proof.* Set $M_\varepsilon = \left\lceil \frac{\mathrm{diam}\,\Omega}{\varepsilon} \right\rceil$, fix $\rho \in \mathcal{R}$ and extend $u$ by 0 to all of $\varepsilon\mathbb{Z}^d$. Then,

$$u(x) = -\varepsilon \sum_{k=1}^{M_\varepsilon} \frac{u(x + k\varepsilon\rho) - u(x + (k-1)\varepsilon\rho)}{\varepsilon}$$

for all $x \in \Omega \cap \varepsilon\mathbb{Z}^d$ and thus

$$\varepsilon^d \sum_{x\in\mathrm{int}_\varepsilon \Omega} |u(x)|^2 \leq \varepsilon^{d+2} \sum_{x\in\mathrm{int}_\varepsilon \Omega} \Big(\sum_{k=1}^{M_\varepsilon}|D_{\mathcal{R},\varepsilon}u(x + (k-1)\varepsilon\rho)|\Big)^2$$

$$\leq \varepsilon^{d+2}M_\varepsilon \sum_{x\in\mathrm{int}_\varepsilon \Omega} \sum_{k=1}^{M_\varepsilon}|D_{\mathcal{R},\varepsilon}u(x + (k-1)\varepsilon\rho)|^2$$

$$\leq \varepsilon^{d+2}M_\varepsilon^2 \sum_{x\in\mathrm{sint}_\varepsilon \Omega} |D_{\mathcal{R},\varepsilon}u(x)|^2$$

$$\leq (\mathrm{diam}\,\Omega + 1)^2\varepsilon^d \sum_{x\in\mathrm{sint}_\varepsilon \Omega} |D_{\mathcal{R},\varepsilon}u(x)|^2.$$

For the second inequality just take a $v \in \mathcal{A}_\varepsilon(\Omega, 0)$ and calculate

$$\varepsilon^d \sum_{x\in\mathrm{int}_\varepsilon \Omega} u(x)v(x) \leq \|u\|_{\ell_\varepsilon^2(\mathrm{int}_\varepsilon \Omega)}\|v\|_{\ell_\varepsilon^2(\mathrm{int}_\varepsilon \Omega)} \leq C_\mathrm{P}\|u\|_{\ell_\varepsilon^2(\mathrm{int}_\varepsilon \Omega)}\|v\|_{h_\varepsilon^1(\mathrm{sint}_\varepsilon \Omega)}$$

$\square$

Now let us prove the theorem.

*Proof of Theorem 4.5.* By Theorem 3.7, $\lambda_{\mathrm{atom}}(A_0) > 0$ implies that also $\lambda_{\mathrm{LH}}(A_0) > 0$ and we can apply Theorem 4.4 with $m = 2$. This already gives a $K_1$ and the solution of the continuous problem $y$. Since the solution depends continuously on the data and we have the embedding $H^4 \hookrightarrow C^1$ we can always achieve

$$|D_{\mathcal{R},\varepsilon}S_\varepsilon y(x) - (A_0\rho)_{\rho\in\mathcal{R}}| \leq \frac{r_0}{2}$$

for all $x \in \varepsilon\mathbb{Z}^d$ and all $\varepsilon \in (0, 1]$ by choosing $K_1$ small enough.

We want to use Corollary 4.8. Let $X_\varepsilon = \mathcal{A}_\varepsilon(\Omega, 0)$ with $\|u\|_{X_\varepsilon} = \|u\|_{h^1_\varepsilon(\mathrm{sint}_\varepsilon \Omega)}$,

$$Z_\varepsilon = \{r \colon \mathrm{int}_\varepsilon \Omega \to \mathbb{R}^d\}$$

with $\|r\|_{Z_\varepsilon} = \|r\|_{h^{-1}_\varepsilon(\mathrm{int}_\varepsilon \Omega)}$ and

$$Y_\varepsilon = \left(\{g \colon \partial_\varepsilon \Omega \to \mathbb{R}^d\} \big/ \{g \colon \|g\|_{\partial_\varepsilon \Omega} = 0\}\right) \times \mathcal{A}_\varepsilon(\Omega, 0),$$

with $\|([g], v)\|_{Y_\varepsilon} = \|g\|_{\partial_\varepsilon \Omega} + \|v\|_{h^{-1}_\varepsilon(\mathrm{int}_\varepsilon \Omega)}$. Note that $D_{\mathcal{R},\varepsilon} T_\varepsilon g(x) = D_{\mathcal{R},\varepsilon} T_\varepsilon h(x)$ for all $x \in \mathrm{sint}_\varepsilon \Omega$ whenever $[g] = [h]$. Now, define $F_\varepsilon \colon X_\varepsilon \times Y_\varepsilon \to Z_\varepsilon$ by

$$\begin{aligned}
F_\varepsilon(u, [g], v)(x) = &-\tilde{f}(x) - v(x) \\
&- \mathrm{div}_{\mathcal{R},\varepsilon}\left(DW_{\mathrm{atom}}(D_{\mathcal{R},\varepsilon}(S_\varepsilon y + T_\varepsilon g + u)(x))\right)
\end{aligned}$$

for $x \in \mathrm{int}_\varepsilon \Omega$. This is well defined for all $\varepsilon \in (0, 1]$ on an open neighborhood of

$$\overline{B_{r_1 \varepsilon^{\frac{d}{2}}}(0)} \times \overline{B_{r_2 \varepsilon^{\frac{d}{2}}}(0)} \times \mathcal{A}_\varepsilon(\Omega, 0),$$

if we choose $r_1, r_2 > 0$ small enough. In particular, we can choose them so small that

$$D_{\mathcal{R},\varepsilon}(S_\varepsilon y + T_\varepsilon g + u)(x) \in \overline{B_{r_0}((A_0 \rho)_{\rho \in \mathcal{R}})}$$

for all $x \in \mathrm{sint}_\varepsilon \Omega$. Now we use Proposition 4.9 with $S_\varepsilon y \in C^{3,1}(\Omega; \mathbb{R}^d)$ and $V = B_{r_0}((A_0 \rho)_{\rho \in \mathcal{R}})$ to get

$$\begin{aligned}
\|F_\varepsilon(0, 0, 0)\|_{\ell^2_\varepsilon(\mathrm{int}_\varepsilon \Omega)} \leq &\|\mathrm{div}\, DW_{\mathrm{CB}}(\nabla y) - \mathrm{div}\, DW_{\mathrm{CB}}(\nabla S_\varepsilon y)\|_{L^2(\Omega;\mathbb{R}^d)} \\
&+ C\varepsilon^2 \Big\| \|\nabla^4 S_\varepsilon y\|_{L^\infty(B_{\varepsilon R}(x))} + \|\nabla^3 S_\varepsilon y\|^{\frac{3}{2}}_{L^\infty(B_{\varepsilon R}(x))} \\
&+ \|\nabla^2 S_\varepsilon y\|^3_{L^\infty(B_{\varepsilon R}(x))} + \varepsilon \|\nabla^3 S_\varepsilon y\|^2_{L^\infty(B_{\varepsilon R}(x))} \Big\|_{L^2(\Omega)}.
\end{aligned}$$

Next, we can apply Proposition 4.10 and Proposition 4.11 on $\bar{y} = y_{A_0} + E(y - y_{A_0})$ and use $y = \bar{y}$ in $\Omega$ to get

$$\begin{aligned}
\|F_\varepsilon(0, 0, 0)\|_{\ell^2_\varepsilon(\mathrm{int}_\varepsilon \Omega)} \leq\ &C\varepsilon^2 \big(\|\nabla^2 \bar{y}\|_{L^4(\mathbb{R}^d)} \|\nabla^3 \bar{y}\|_{L^4(\mathbb{R}^d)} + \|\nabla^4 \bar{y}\|_{L^2(\mathbb{R}^d)} \\
&+ \|\nabla^3 \bar{y}\|^{\frac{3}{2}}_{L^3(\mathbb{R}^d)} + \|\nabla^2 \bar{y}\|^3_{L^6(\mathbb{R}^d)} + \varepsilon \|\nabla^3 \bar{y}\|^2_{L^4(\mathbb{R}^d)}\big) \\
\leq\ &C_1 \varepsilon^2 \|y - y_{A_0}\|_{H^4(\Omega;\mathbb{R}^d)} (1 + \|y - y_{A_0}\|^2_{H^4(\Omega;\mathbb{R}^d)})
\end{aligned}$$

Hence, we can set

$$A = C_{\mathrm{P}} C_1 \|y - y_{A_0}\|_{H^4(\Omega;\mathbb{R}^d)} (1 + \|y - y_{A_0}\|^2_{H^4(\Omega;\mathbb{R}^d)}).$$

By stability,

$$\varepsilon^d \sum_{x \in \mathrm{sint}_\varepsilon \Omega} D^2 W_{\mathrm{atom}}((A_0\rho)_{\rho \in \mathcal{R}})[D_{\mathcal{R},\varepsilon} u(x)]^2$$

$$\geq \lambda_{\mathrm{atom}}(A_0)\varepsilon^d \sum_{x \in \mathrm{sint}_\varepsilon \Omega} |D_{\mathcal{R},\varepsilon} u(x)|^2$$

for all $u \in \mathcal{A}_\varepsilon(\Omega, 0)$. Continuity of $D^2 W_{\mathrm{atom}}$ then implies the existence of a $\tilde{r} \leq r_0$ such that

$$\varepsilon^d \sum_{x \in \mathrm{sint}_\varepsilon \Omega} D^2 W_{\mathrm{atom}}(D_{\mathcal{R},\varepsilon} w(x))[D_{\mathcal{R},\varepsilon} u(x)]^2$$

$$\geq \frac{\lambda_{\mathrm{atom}}(A_0)}{2}\varepsilon^d \sum_{x \in \mathrm{sint}_\varepsilon \Omega} |D_{\mathcal{R},\varepsilon} u(x)|^2$$

for all $u \in \mathcal{A}_\varepsilon(\Omega, 0)$ and all $w \colon \Omega \cap \varepsilon\mathbb{Z}^d \to \mathbb{R}^d$ with

$$|D_{\mathcal{R},\varepsilon} w(x) - (A_0\rho)_{\rho \in \mathcal{R}}| \leq \tilde{r}$$

for all $x \in \mathrm{sint}_\varepsilon \Omega$. And again, by choosing $K_1$ small enough this last inequality is automatically satisfied for $w = S_\varepsilon y$ with $\varepsilon \in (0, 1]$ arbitrary.

Since the spaces are finite dimensional it is obvious that the $F_\varepsilon$ are Fréchet-differentiable. For $w \in X_\varepsilon$ we have

$$\varepsilon^d \sum_{x \in \mathrm{int}_\varepsilon \Omega} \partial_u F_\varepsilon(0, 0, 0)[w](x) w(x)$$

$$= -\varepsilon^{d-1} \sum_{x \in \mathrm{int}_\varepsilon \Omega} \sum_{\sigma, \rho} w(x) \Big( D_{e_\rho} D_{e_\sigma} W_{\mathrm{atom}}(D_{\mathcal{R},\varepsilon} S_\varepsilon y(x)) \frac{w(x + \varepsilon\sigma) - w(x)}{\varepsilon}$$

$$- D_{e_\rho} D_{e_\sigma} W_{\mathrm{atom}}(D_{\mathcal{R},\varepsilon} S_\varepsilon y(x - \varepsilon\rho)) \frac{w(x + \varepsilon\sigma - \varepsilon\rho) - w(x - \varepsilon\rho)}{\varepsilon} \Big)$$

$$= \varepsilon^d \sum_{x \in \mathrm{sint}_\varepsilon \Omega} D^2 W_{\mathrm{atom}}(D_{\mathcal{R},\varepsilon} S_\varepsilon y(x))[D_{\mathcal{R},\varepsilon} w(x), D_{\mathcal{R},\varepsilon} w(x)]$$

$$\geq \frac{\lambda_{\mathrm{atom}}(A_0)}{2}\varepsilon^d \sum_{x \in \mathrm{sint}_\varepsilon \Omega} |D_{\mathcal{R},\varepsilon} w(x)|^2$$

and, thus,

$$\|\partial_u F_\varepsilon(0, 0, 0)^{-1}\|_{L(h_\varepsilon^{-1}(\mathrm{int}_\varepsilon \Omega), h_\varepsilon^1(\mathrm{sint}_\varepsilon \Omega))} \leq \frac{2}{\lambda_{\mathrm{atom}}(A_0)} = M_1.$$

Furthermore, for $([h], v) \in Y_\varepsilon$ and $w \in \mathcal{A}_\varepsilon(\Omega, 0)$ we have

$$
\varepsilon^d \sum_{x \in \mathrm{int}_\varepsilon \Omega} \partial_{([g], v)} F_\varepsilon(0, 0, 0)[([h], v)](x) w(x)
$$

$$
\leq \|v\|_{h_\varepsilon^{-1}(\mathrm{int}_\varepsilon \Omega)} \|w\|_{h_\varepsilon^1(\mathrm{sint}_\varepsilon \Omega)}
$$

$$
- \varepsilon^{d-1} \sum_{x \in \mathrm{int}_\varepsilon \Omega} \sum_{\sigma, \rho \in \mathcal{R}} w(x)
$$

$$
\cdot \left( D_{e_\rho} D_{e_\sigma} W_{\mathrm{atom}}(D_{\mathcal{R}, \varepsilon} S_\varepsilon y(x)) \frac{T_\varepsilon h(x + \varepsilon \sigma) - T_\varepsilon h(x)}{\varepsilon} \right.
$$

$$
\left. - D_{e_\rho} D_{e_\sigma} W_{\mathrm{atom}}(D_{\mathcal{R}, \varepsilon} S_\varepsilon y(x - \varepsilon \rho)) \frac{T_\varepsilon h(x - \varepsilon \rho + \varepsilon \sigma) - T_\varepsilon h(x - \varepsilon \rho)}{\varepsilon} \right)
$$

$$
= \|v\|_{h_\varepsilon^{-1}(\mathrm{int}_\varepsilon \Omega)} \|w\|_{h_\varepsilon^1(\mathrm{sint}_\varepsilon \Omega)}
$$

$$
+ \varepsilon^d \sum_{x \in \mathrm{sint}_\varepsilon \Omega} D^2 W_{\mathrm{atom}}(D_{\mathcal{R}, \varepsilon} S_\varepsilon y(x))[D_{\mathcal{R}, \varepsilon} w(x), D_{\mathcal{R}, \varepsilon} T_\varepsilon h(x)]
$$

$$
\leq \|v\|_{h_\varepsilon^{-1}(\mathrm{int}_\varepsilon \Omega)} \|w\|_{h_\varepsilon^1(\mathrm{sint}_\varepsilon \Omega)} + \|D^2 W_{\mathrm{atom}}\|_\infty \|w\|_{h_\varepsilon^1(\mathrm{sint}_\varepsilon \Omega)} \|h\|_{\partial_\varepsilon \Omega}.
$$

Hence,

$$
\|\partial_{([g], v)} F_\varepsilon(0, 0, 0)\|_{L(Y_\varepsilon, Z_\varepsilon)} \leq 1 + \|D^2 W_{\mathrm{atom}}\|_\infty = M_2.
$$

In a similar fashion we calculate

$$
\varepsilon^d \sum_{x \in \mathrm{int}_\varepsilon \Omega} \left( \partial_u F_\varepsilon(0, 0, 0) - \partial_u F_\varepsilon(u, [g], v) \right)[w](x) z(x)
$$

$$
= \varepsilon^d \sum_{x \in \mathrm{sint}_\varepsilon \Omega} \left( D^2 W_{\mathrm{atom}}(D_{\mathcal{R}, \varepsilon} S_\varepsilon y(x)) \right.
$$

$$
\left. - D^2 W_{\mathrm{atom}}(D_{\mathcal{R}, \varepsilon}(S_\varepsilon y + u + T_\varepsilon g)(x))) [D_{\mathcal{R}, \varepsilon} w(x), D_{\mathcal{R}, \varepsilon} z(x)] \right.
$$

$$
\leq \|w\|_{h_\varepsilon^1(\mathrm{sint}_\varepsilon \Omega)} \|z\|_{h_\varepsilon^1(\mathrm{sint}_\varepsilon \Omega)} \|D^3 W_{\mathrm{atom}}\|_\infty \|D_{\mathcal{R}, \varepsilon}(u + T_\varepsilon g)\|_{\ell^\infty(\mathrm{sint}_\varepsilon \Omega)}.
$$

Thus,

$$
\|\partial_u F_\varepsilon(0, 0, 0) - \partial_u F_\varepsilon(u, [g], v)\|_{L(X_\varepsilon, Z_\varepsilon)}
$$

$$
\leq \|D^3 W_{\mathrm{atom}}\|_\infty \varepsilon^{-\frac{d}{2}} (\|u\|_{h_\varepsilon^1(\mathrm{sint}_\varepsilon \Omega)} + \|g\|_{\partial_\varepsilon \Omega}),
$$

so that we can take $M_3 = \|D^3 W_{\mathrm{atom}}\|_\infty$.

At last,

$$\varepsilon^d \sum_{x \in \text{int}_\varepsilon \Omega} \big(\partial_{([g],v)} F_\varepsilon(0,0,0) - \partial_{([g],v)} F_\varepsilon(u,[g],v)\big)[([h],w)](x)z(x)$$

$$= \varepsilon^d \sum_{x \in \text{sint}_\varepsilon \Omega} \big(D^2 W_{\text{atom}}(D_{\mathcal{R},\varepsilon} S_\varepsilon y(x))$$

$$- D^2 W_{\text{atom}}(D_{\mathcal{R},\varepsilon}(S_\varepsilon y + u + T_\varepsilon g)(x)))\big)[D_{\mathcal{R},\varepsilon} T_\varepsilon h(x), D_{\mathcal{R},\varepsilon} z(x)]$$

$$\leq \|h\|_{\partial_\varepsilon \Omega} \|z\|_{h^1_\varepsilon(\text{sint}_\varepsilon \Omega)} \|D^3 W_{\text{atom}}\|_\infty \|D_{\mathcal{R},\varepsilon}(u + T_\varepsilon g)\|_{\ell^\infty(\text{sint}_\varepsilon \Omega)}.$$

Hence, we can also take $M_4 = \|D^3 W_{\text{atom}}\|_\infty$. As before, since $y$ depends continuously on the data, we can take $K_1$ small enough such that

$$C_{\text{P}} C_1 \|y - y_{A_0}\|_{H^4(\Omega;\mathbb{R}^d)}(1 + \|y - y_{A_0}\|_{H^4(\Omega;\mathbb{R}^d)}^2)$$

$$\leq \min\left\{ \frac{r_1 \lambda_{\text{atom}}(A_0)}{8}, \frac{\lambda_{\text{atom}}(A_0)^2}{64 \|D^3 W_{\text{atom}}\|_\infty} \right\}.$$

Therefore, we can apply Corollary 4.8 and get the fixed point result with

$$\lambda_1 = \min\left\{ r_1, \frac{\lambda_{\text{atom}}(A_0)}{8 \|D^3 W_{\text{atom}}\|_\infty} \right\},$$

$$\lambda_2 = \min\left\{ r_2, \frac{\lambda_{\text{atom}}(A_0)}{8 \|D^3 W_{\text{atom}}\|_\infty}, \frac{r_1 \lambda_{\text{atom}}(A_0)}{8(1 + \|D^2 W_{\text{atom}}\|_\infty) + 2\lambda_{\text{atom}}(A_0)} \right.$$

$$\left. \frac{\lambda_{\text{atom}}(A_0)^2}{8 \|D^3 W_{\text{atom}}\|_\infty \big(8 + 8\|D^2 W_{\text{atom}}\|_\infty + 2\lambda_{\text{atom}}(A_0)\big)} \right\}.$$

After doing the substitutions $g_{\text{atom}} \in S_\varepsilon y + [g]$, $f_{\text{atom}} = \tilde{f} + v$ and $y_{\text{atom}} = S_\varepsilon y + T_\varepsilon(g_{\text{atom}} - S_\varepsilon y) + u$, we get the stated existence result with $K_2 = \frac{\lambda_2}{2}$. The solution then satisfies $\|y_{\text{atom}} - S_\varepsilon y\|_{h^1_\varepsilon(\text{sint}_\varepsilon \Omega)} \leq K_3 \varepsilon^\gamma$ with $K_3 = \lambda_1 + \frac{\lambda_2}{2}$. If $r_1, r_2$ are chosen small enough then $\|\tilde{y}_{\text{atom}} - S_\varepsilon y\|_{h^1_\varepsilon(\text{sint}_\varepsilon \Omega)} \leq K_3 \varepsilon^\gamma$ implies

$$|D_{\mathcal{R},\varepsilon} \tilde{y}_{\text{atom}}(x) - (A_0 \rho)_{\rho \in \mathcal{R}}| \leq \frac{\tilde{r}}{2}$$

for all $x \in \text{sint}_\varepsilon \Omega$ and any $\tilde{y}_{\text{atom}}$ with boundary values $g_{\text{atom}}$. Furthermore, with $r_1, r_2$ chosen small enough for $u \in \mathcal{A}_\varepsilon(\Omega, 0) \backslash \{0\}$ with $\|u\|_{h^1_\varepsilon(\text{sint}_\varepsilon \Omega)} \leq K_3 \varepsilon^\gamma$ we have

$$|D_{\mathcal{R},\varepsilon} u(x)| \leq \frac{\tilde{r}}{2}.$$

Now, since $y_{\text{atom}}$ is a solution, we can calculate

$$E_\varepsilon(y_{\text{atom}} + u, f_{\text{atom}}, g_{\text{atom}}) - E_\varepsilon(y_{\text{atom}}, f_{\text{atom}}, g_{\text{atom}})$$

$$= \varepsilon^d \sum_{x \in \text{sint}_\varepsilon \Omega} \Big( W_{\text{atom}}(D_{\mathcal{R},\varepsilon} y_{\text{atom}}(x) + D_{\mathcal{R},\varepsilon} u(x)) - W_{\text{atom}}(D_{\mathcal{R},\varepsilon} y_{\text{atom}}(x))$$

$$- DW_{\text{atom}}(D_{\mathcal{R},\varepsilon} y_{\text{atom}}(x))[D_{\mathcal{R},\varepsilon} u(x)] \Big)$$

$$= \varepsilon^d \sum_{x \in \text{sint}_\varepsilon \Omega} \int_0^1 (1-t) D^2 W_{\text{atom}}(D_{\mathcal{R},\varepsilon} y_{\text{atom}}(x) + t D_{\mathcal{R},\varepsilon} u(x))$$

$$[D_{\mathcal{R},\varepsilon} u(x), D_{\mathcal{R},\varepsilon} u(x)] \, dt$$

$$\geq \frac{\lambda_{\text{atom}}(A_0)}{2} \varepsilon^d \sum_{x \in \text{sint}_\varepsilon \Omega} |D_{\mathcal{R},\varepsilon} u(x)|^2 > 0,$$

which shows that $y_{\text{atom}}$ is a strict local minimizer. And, doing the same calculation again with $\tilde{y}_{\text{atom}} - y_{\text{atom}}$ instead of $u$, we also see that the solution is unique.

For the additional statement we only have to estimate the quantity $\|S_\varepsilon y - y\|_{h^1_\varepsilon(\text{int}_\varepsilon \Omega)}$ with a Taylor expansion. Indeed, we have

$$\left| \frac{S_\varepsilon y(x + \varepsilon \rho) - S_\varepsilon y(x)}{\varepsilon} - \frac{y(x + \varepsilon \rho) - y(x)}{\varepsilon} \right|^2$$

$$\leq \left| \int_{B_\varepsilon(0)} \frac{\bar{y}(x + \varepsilon \rho + z) - \bar{y}(x + z) - \bar{y}(x + \varepsilon \rho) + \bar{y}(x)}{\varepsilon} \eta_\varepsilon(z) \, dz \right|^2$$

$$\leq \left| \int_{B_\varepsilon(0)} \Big( \frac{\bar{y}(x + \varepsilon \rho + z) - \bar{y}(x + \varepsilon \rho) - \nabla \bar{y}(x + \varepsilon \rho)[z]}{\varepsilon} \right.$$

$$\left. - \frac{\bar{y}(x + z) - \bar{y}(x) - \nabla \bar{y}(x)[z]}{\varepsilon} \Big) \eta_\varepsilon(z) \, dz \right|^2$$

$$\leq \varepsilon^{-2} \left| \int_{B_\varepsilon(0)} \int_0^1 (1-t) \big( \nabla^2 \bar{y}(x + \varepsilon \rho + tz)[z, z] \right.$$

$$\left. - \nabla^2 \bar{y}(x + tz)[z, z] \big) \eta_\varepsilon(z) \, dt \, dz \right|^2$$

$$\leq \varepsilon^{2\gamma} R_{\max}^{2(\gamma-1)} |\nabla^2 \bar{y}|_{\gamma-1}^2,$$

which gives the desired result.  $\square$

# Chapter 5

# The Dynamic Case

## 5.1   Continuum Elastodynamics

Before we can discuss the atomistic equations, we have to discuss the continuous Cauchy-Born problem. We are not only interested in existence and uniqueness, but also in the maximal existence interval and higher order regularity. The most important part of the result is the short time existence, already contained in [DH85]. But, since we want to discuss the atomistic equations for long times, it is important that extend this short time result to a result about the maximal existence time. This will still require a considerable amount of work. Furthermore, there are two key regularity theorems that are only stated and used in [DH85]. Their proofs were left out by the authors. Since we will also need these statements in our proof directly, we will prove them in the appendix. Theorem A.2 is about additional regularity for solutions of second order hyperbolic equations, while Theorem C.3 is about higher order elliptic regularity under very weak assumptions on the coefficients.

### The Compatibility Conditions

If we want to achieve higher regularity for such a second order hyperbolic initial-boundary-value problem, compatibility conditions on $f, g, h_0, h_1$ are crucial. We say that $f, g, h_0, h_1$ satisfy the compatibility conditions of order $m$, if

$$u_k := \frac{\partial^k}{\partial t^k}(y - g)|_{t=0} \in H_0^1(\Omega; \mathbb{R}^d)$$

for all $k \in \{0, \ldots, m-1\}$ as computed formally using the equation in terms of $f, g, h_0, h_1$. This can be written explicitly and, even though the

117

expressions are quite nasty, we still want to do so in order to be able to discuss some regularity issues in more detail.

If $m - 1 > \frac{d}{2}$, $h_0 \in H^m(\Omega; \mathbb{R}^d)$, $h_1 \in H^{m-1}(\Omega; \mathbb{R}^d)$, $\partial_t^{k-2} f(\cdot, 0) \in H^{m-k}(\Omega; \mathbb{R}^d)$ for $2 \le k \le m - 1$, $\partial_t^k g(\cdot, 0) \in H^{m-k}(\Omega; \mathbb{R}^d)$ for $0 \le k \le m - 1$ and $W_{\mathrm{CB}} \in C^{m-1}$ on an open set containing $\{\nabla h_0(x) \colon x \in \Omega\}$, then we define $u_0(x) = h_0(x) - g(x, 0)$, $u_1(x) = h_1(x) - \dot{g}(x, 0)$,

$$u_2(x) = f(x, 0) - \ddot{g}(x, 0) + \operatorname{div}\big(DW_{\mathrm{CB}}(\nabla h_0(x))\big)$$

and recursively, for $3 \le k \le m - 1$,

$$(u_k(x))_i = \partial_t^{k-2} f_i(x, 0) - \partial_t^k g_i(x, 0)$$

$$+ \sum_{j,q,r=1}^d D^{E_{ij}+E_{qr}} W_{\mathrm{CB}}(\nabla h_0(x))\Big(\partial_{x_r}\partial_{x_j}\partial_t^{k-2} g_q(x, 0) + \partial_{x_r}\partial_{x_j}(u_{k-2})_q(x)\Big)$$

$$+ \sum_{j,q,r=1}^d \sum_{n=1}^{k-2} \sum_{\substack{\beta \in \mathbb{N}_0^{d\times d} \\ 1 \le |\beta| \le n}} \sum_{s=1}^n \sum_{p_s(n,\beta)} \frac{(k-2)!}{(k-2-n)!} D^{\beta+E_{ij}+E_{qr}} W_{\mathrm{CB}}(\nabla h_0(x))$$

$$\cdot \Big(\prod_{l=1}^s \frac{(\partial_t^{\gamma_l}\nabla g(x, 0) + \nabla u_{\gamma_l}(x))^{\lambda_l}}{\lambda_l!(\gamma_l!)^{\lambda_l}}\Big)$$

$$\cdot \Big(\partial_{x_r}\partial_{x_l}\partial_t^{k-2-n} g_q(x, 0) + \partial_{x_r}\partial_{x_l}(u_{k-2-n})_q(x)\Big),$$

where

$$p_s(n, \beta) = \Big\{(\lambda_1, \ldots, \lambda_s; \gamma_1, \ldots, \gamma_s) \colon \lambda_l \in \mathbb{N}_0^{d\times d}, \gamma_l \in \mathbb{N}_0,$$

$$0 < \gamma_1 < \cdots < \gamma_s, |\lambda_l| > 0, \sum_{l=1}^s \lambda_l = \beta, \sum_{l=1}^s \gamma_l|\lambda_l| = n\Big\},$$

and $E_{ij}$ is the matrix with $(E_{ij})_{ij} = 1$ and zeros everywhere else. The following result shows that $u_k \in H_0^1(\Omega; \mathbb{R}^d)$ is only a condition on the boundary values and not on the regularity.

**Proposition 5.1.** *If $m - 1 > \frac{d}{2}$, $h_0 \in H^m(\Omega; \mathbb{R}^d)$, $h_1 \in H^{m-1}(\Omega; \mathbb{R}^d)$, $\partial_t^{k-2} f(\cdot, 0) \in H^{m-k}(\Omega; \mathbb{R}^d)$ for $2 \le k \le m-1$, $\partial_t^k g(\cdot, 0) \in H^{m-k}(\Omega; \mathbb{R}^d)$ for $0 \le k \le m - 1$ and $W_{\mathrm{CB}} \in C^m$ on an open set containing*

$$\{\nabla h_0(x) \colon x \in \Omega\},$$

*then $u_k \in H^{m-k}(\Omega; \mathbb{R}^d)$ for all $0 \le k \le m - 1$.*

*Proof.* This is clear for $k = 0, 1$. For $k = 2$ this follows directly from Lemma B.5. If $k \geq 3$, we find $D^{\beta + E_{ij} + E_{qr}} W_{\text{CB}} \circ \nabla h_0 \in H^{m-1}$ and, by induction, can then apply Lemma B.1 with $M = m - 1$ and $N = 0 + \sum \gamma_j |\lambda_j| + (k - l - 1) = k - 1$ to estimate the product and get the desired result. Actually, for this to be completely true, we would need the stronger assumption $W_{\text{CB}} \in C^{2m-2}$ so that Lemma B.5 gives $D^{\beta + E_{ij} + E_{qr}} W_{\text{CB}} \circ \nabla h_0 \in H^{m-1}$. To reduce this assumption to $W_{\text{CB}} \in C^m$, we note that in the application of Lemma B.1 we only take the $\alpha$-th derivative of $v = D^{\beta + E_{ij} + E_{qr}} W_{\text{CB}} \circ \nabla h_0$ with $0 \leq |\alpha| \leq m - k = M - N$ and then have to know that $D^{\alpha} v \in L^q$ for a certain $q$ formerly coming from the Sobolev embedding of $H^{m-1-|\alpha|}$. Now we have to prove this estimate differently. From Corollary B.4 we know that

$$|D^{\alpha} v(x)| \leq C \sum_{r=1}^{|\alpha|} |D^{r+2+|\beta|} W_{\text{CB}}(\nabla h_0(x))| \sum_{\substack{l_1, \ldots, l_r \geq 1 \\ l_1 + \cdots + l_r = |\alpha|}} \prod_{j=1}^{r} |D^{(l_j+1)} h_0(x)|,$$

if $W_{\text{CB}} \in C^m$ and $h_0 \in C^{1+|\alpha|}$. But of course this extends to $h_0 \in H^m$ once we estimate the product on the right hand side suitably. These estimates, which also give the desired integrability of $D^{\alpha} v$ follow along the lines of the proof of Lemma B.1. $\square$

If we already have a solution and use it as a starting point, then the compatibility conditions are automatically satisfied and the $u_k$ are indeed directly given by $y - g$.

**Proposition 5.2.** *Let $m \in \mathbb{N}$ with $m > \frac{d}{2} + 1$, $\delta > 0$, let $\Omega \subset \mathbb{R}^d$ be an open, bounded set with $\partial \Omega$ of class $C^m$, $V \subset \mathbb{R}^{d \times d}$ open, $W_{\text{CB}} \in C_b^{m+1}(V)$,*

$$f \in C^{m-1}(\overline{\Omega} \times [-\delta, \delta]; \mathbb{R}^d),$$
$$g \in C^{m+1}(\overline{\Omega} \times [-\delta, \delta]; \mathbb{R}^d), \text{ and}$$
$$y \in \bigcap_{k=0}^{m} C^k([-\delta, \delta]; H^{m-k}(\Omega; \mathbb{R}^d)) \text{ with}$$
$$\overline{\{\nabla y(x, t) \colon x \in \Omega, t \in [-\delta, \delta]\}} \subset V_{\text{LH}}.$$

*Furthermore let $y$ be a solution of the equations. If we now set $h_0 = y(0)$ and $h_1 = \partial_t y(0)$, then we have*

$$u_k = \frac{\partial^k}{\partial t^k}(y - g)|_{t=0} \in H_0^1(\Omega; \mathbb{R}^d)$$

*for all $k \in \{0, \ldots, m-1\}$.*

*Proof.* Since $y - g \in C^{m-1}\big([-\delta, \delta]; H^1(\Omega; \mathbb{R}^d)\big)$, $H_0^1(\Omega; \mathbb{R}^d)$ is a closed subspace of $H^1(\Omega; \mathbb{R}^d)$ and $y(t) - g(t) \in H_0^1(\Omega; \mathbb{R}^d)$ for all $t \in [-\delta, \delta]$, we clearly find

$$\frac{\partial^k}{\partial t^k}(y - g)|_{t=0} \in H_0^1(\Omega; \mathbb{R}^d).$$

Let us now proof $u_k = \frac{\partial^k}{\partial t^k}(y - g)|_{t=0}$ by induction over $k$. By definition this is true for $k = 0, 1$. $k = 2$ follows from the equation. If now $3 \leq k \leq m - 1$, we have to show that the recursion formula for the $u_k$ also holds for the derivatives of $y - g$. Clearly we have

$$\partial_t^k(y - g) = \partial_t^{k-2}\Big(f - \partial_t^2 g + \operatorname{div}(DW_{\mathrm{CB}}(\nabla y))\Big)$$
$$= \partial_t^{k-2}f - \partial_t^k g + \partial_t^{k-2}\operatorname{div}(DW_{\mathrm{CB}}(\nabla y)).$$

If $y$ were smooth the last term can be written explicitly with the chain rule, the Leibniz rule, as well as the generalized Faà di Bruno formula, Lemma B.3. We first get

$$\operatorname{div}(DW_{\mathrm{CB}}(\nabla y))_i = \sum_{j,q,r=1}^{d} D^{E_{ij}+E_{qr}}W_{\mathrm{CB}}(\nabla y)\partial_{x_r}\partial_{x_j}y_q,$$

then

$$\partial_t^{k-2}\operatorname{div}(DW_{\mathrm{CB}}(\nabla y))_i$$
$$= \sum_{j,q,r=1}^{d}\sum_{n=0}^{k-2}\binom{k-2}{n}\partial_t^n(D^{E_{ij}+E_{qr}}W_{\mathrm{CB}}(\nabla y))\partial_t^{k-2-n}\partial_{x_r}\partial_{x_j}y_q,$$

and finally

$$
\begin{aligned}
&(\partial_t^k(y-g))_i \\
&= \partial_t^{k-2} f_i - \partial_t^k g_i \\
&\quad + \sum_{j,q,r=1}^d D^{E_{ij}+E_{qr}} W_{\mathrm{CB}}(\nabla y) \partial_{x_r} \partial_{x_j} \partial_t^{k-2} y_q \\
&\quad + \sum_{j,q,r=1}^d \sum_{n=1}^{k-2} \sum_{\substack{\beta \in \mathbb{N}_0^{d\times d} \\ 1 \le |\beta| \le n}} \sum_{s=1}^n \sum_{p_s(n,\beta)} \frac{(k-2)!}{(k-2-n)!} D^{\beta+E_{ij}+E_{qr}} W_{\mathrm{CB}}(\nabla y) \\
&\qquad \cdot \Big( \prod_{l=1}^s \frac{(\partial_t^{\gamma_l} \nabla y)^{\lambda_l}}{\lambda_l! (\gamma_l!)^{|\lambda_l|}} \Big) \partial_{x_r} \partial_{x_l} \partial_t^{k-2-n} y_q,
\end{aligned}
$$

where

$$
\begin{aligned}
p_s(n,\beta) = \Big\{ (\lambda_1,\dots,\lambda_s;\gamma_1,\dots,\gamma_s) \colon \lambda_l \in \mathbb{N}_0^{d\times d}, \gamma_l \in \mathbb{N}_0, \\
0 < \gamma_1 < \cdots < \gamma_s, |\lambda_l| > 0, \sum_{l=1}^s \lambda_l = \beta, \sum_{l=1}^s \gamma_l |\lambda_l| = n \Big\}.
\end{aligned}
$$

Due to the bounds discussed in Proposition 5.1 and Lemma B.1 this still holds under the given weaker regularity assumption on $y$. Inductively, we thus have proven the claim. $\qquad\square$

## Local Existence

In the following, for $V \subset \mathbb{R}^{d\times d}$ open and $W_{\mathrm{CB}} \in C^2(V)$ we write

$$
V_{\mathrm{LH}} = \{ A \in V \colon \lambda_{\mathrm{LH}}(A) > 0 \},
$$

which is again an open set, since $\lambda_{\mathrm{LH}}$ is continuous.

Let us start with a local existence result.

**Theorem 5.3.** *Let $m \in \mathbb{N}$ with $m > \frac{d}{2} + 2$, $T_0 > 0$, let $\Omega \subset \mathbb{R}^d$ be an open, bounded set with $\partial\Omega$ of class $C^m$, $V \subset \mathbb{R}^{d\times d}$ open and $W_{\mathrm{CB}} \in C_b^{m+1}(V)$. Given a body force $f$, initial data $h_0, h_1$ and boundary* *values g such that*

$$f \in C^{m-1}(\overline{\Omega} \times [0, T_0]; \mathbb{R}^d)$$
$$g \in C^{m+1}(\overline{\Omega} \times [0, T_0]; \mathbb{R}^d)$$
$$h_0 \in H^m(\Omega; \mathbb{R}^d)$$
$$\overline{\{\nabla h_0(x) \colon x \in \Omega\}} \subset V_{\mathrm{LH}}$$
$$h_1 \in H^{m-1}(\Omega; \mathbb{R}^d)$$

*and such that the compatibility conditions of order m are satisfied (see above).*

*Then, for all sufficiently small $T \in (0, T_0]$ the problem has a unique solution*

$$y \in \bigcap_{k=0}^{m} C^k([0, T]; H^{m-k}(\Omega; \mathbb{R}^d))$$

*and we have*

$$\overline{\{\nabla y(x, t) \colon x \in \Omega, t \in [0, T]\}} \subset V_{\mathrm{LH}}.$$

*Proof.* This follows from [DH85, Thm. 5.1] by setting $\mathbf{u}^0 = h_0 - g(\cdot, 0)$, $\mathbf{u}^1 = h_1 - \dot{g}(\cdot, 0)$,

$$\mathbf{g}(x, t, u, p, M)_k = f(x, t)_k - \ddot{g}(x, t)_k$$
$$+ \sum \frac{\partial^2 W_{\mathrm{CB}}}{\partial a_{ki} \partial a_{lj}}(M + \nabla g(x, t)) \frac{\partial^2 g_l}{\partial x_i \partial x_j}(x, t)$$

and

$$(\mathbf{A}_{ij})_{kl}(x, t, u, p, M) = \chi(M + \nabla g(x, t)) \frac{\partial^2 W_{\mathrm{CB}}}{\partial a_{ki} \partial a_{lj}}(M + \nabla g(x, t))$$
$$+ (1 - \chi(M + \nabla g(x, t)))\delta_{kl}\delta_{ij}$$

where $\chi \colon \mathbb{R}^{d \times d} \to [0, 1]$ is a smooth cutoff with $\chi(M) = 1$ for $M \in W_1$ and $\chi(M) = 0$ for $M \notin W_2$ and $W_1, W_2$ are open sets such that

$$\overline{\{\nabla h_0(x) \colon x \in \Omega\}} \subset\subset W_1 \subset\subset W_2 \subset\subset V_{\mathrm{LH}}.$$

We then set $y = u + g$ and reduce the existence time $T$ enough to ensure $\nabla y(x, t) \in W_1$ for all $(x, t) \in \overline{\Omega} \times [0, T]$.                                   $\square$

## The Maximal Existence Time of Solutions

We can use the local result to construct a maximal solution.

**Theorem 5.4.** *Let* $m \in \mathbb{N}$ *with* $m > \frac{d}{2} + 2$, $T_0 > 0$, *let* $\Omega \subset \mathbb{R}^d$ *be an open, bounded set with* $\partial\Omega$ *of class* $C^m$, $V \subset \mathbb{R}^{d \times d}$ *open and* $W_{\mathrm{CB}} \in C^{m+1}(V)$. *Given a body force* $f$, *initial data* $h_0, h_1$ *and boundary values* $g$ *such that*

$$f \in C^{m-1}(\overline{\Omega} \times [0, T_0]; \mathbb{R}^d)$$
$$g \in C^{m+1}(\overline{\Omega} \times [0, T_0]; \mathbb{R}^d)$$
$$h_0 \in H^m(\Omega; \mathbb{R}^d)$$
$$\overline{\{\nabla h_0(x) \colon x \in \Omega\}} \subset V_{\mathrm{LH}}$$
$$h_1 \in H^{m-1}(\Omega; \mathbb{R}^d)$$

*and such that the compatibility conditions of order* $m$ *are satisfied. Then there are unique* $T_{\mathrm{cont}} > 0$ *and*

$$y \in \bigcap_{k=0}^{m} C^k\big([0, T_{\mathrm{cont}}); H^{m-k}(\Omega; \mathbb{R}^d)\big),$$

*such that*
$$\{\nabla y(x, t) \colon x \in \overline{\Omega}, t \in [0, T_{\mathrm{cont}})\} \subset V_{\mathrm{LH}},$$

$y$ *is a solution on* $[0, T_{\mathrm{cont}})$ *and at least one of the following conditions is true:*

*(i)* $T_{\mathrm{cont}} = T_0$,

*(ii)* $\liminf_{t \to T_{\mathrm{cont}}} \mathrm{dist}\big(V_{\mathrm{LH}}^{\mathrm{c}}, \{\nabla y(x, t) \colon x \in \overline{\Omega}\}\big) = 0$,

*(iii)* $\limsup_{t \to T_{\mathrm{cont}}} \|y(t)\|_{H^m(\Omega; \mathbb{R}^d)} + \|\partial_t y(t)\|_{H^{m-1}(\Omega; \mathbb{R}^d)} = \infty$.

*Proof.* Let $T_{\mathrm{cont}}$ be the supremum of all $0 < T \le T_0$ such that there is a

$$y \in \bigcap_{k=0}^{m} C^k\big([0, T]; H^{m-k}(\Omega; \mathbb{R}^d)\big)$$

with
$$\overline{\{\nabla y(x, t) \colon x \in \Omega, t \in [0, T]\}} \subset V_{\mathrm{LH}}$$

that solves the problem on $[0, T]$. Theorem 5.3 ensures that there is at least one such $T$. If we take any of these solutions and $t_0 \in (0, T)$ then

$h_0 = y(t_0)$, $h_1 = \partial_t y(t_0)$ as well as the translated $f$ and $g$ can be used in Theorem 5.3 to show existence and uniqueness in some $[t_0, t_0+\delta]$, $\delta > 0$. This is possible since the new $u_k$ satisfies $u_k = \frac{\partial^k}{\partial t^k}(y - g)|_{t=t_0} \in H_0^1$ by Proposition 5.2.

The uniqueness ensures in particular, that all these $y$ are equal pairwise on the intersection of their existence intervals. Therefore, we have a

$$y \in \bigcap_{k=0}^{m} C^k\big([0, T_{\text{cont}}); H^{m-k}(\Omega; \mathbb{R}^d)\big)$$

with

$$\{\nabla y(x, t) \colon x \in \overline{\Omega}, t \in [0, T_{\text{cont}})\} \subset V_{\text{LH}}$$

that solves the problem on $[0, T_{\text{cont}})$.

Now assume $T_{\text{cont}} < T_0$,

$$\liminf_{t \to T_{\text{cont}}} \text{dist}\big(V_{\text{LH}}^c, \{\nabla y(x, t) \colon x \in \overline{\Omega}\}\big) > 0$$

and

$$\limsup_{t \to T_{\text{cont}}} \|y(t)\|_{H^m(\Omega; \mathbb{R}^d)} + \|\partial_t y(t)\|_{H^{m-1}(\Omega; \mathbb{R}^d)} < \infty.$$

Then

$$y \in L^\infty(0, T_{\text{cont}}; H^m(\Omega; \mathbb{R}^d)) \cap W^{1,\infty}(0, T_{\text{cont}}; H^{m-1}(\Omega; \mathbb{R}^d)).$$

We claim that

$$\partial_t^k y \in L^\infty(0, T_{\text{cont}}; H^{m-k}(\Omega; \mathbb{R}^d))$$

for $0 \le k \le m$. We already know this $k = 0, 1$. For $2 \le k \le m$ we represent the derivatives of $y$ as we did in Proposition 5.2 and then argue inductively as in the proof of Proposition 5.1.

In particular, the limit $\tilde{h}_k := \lim_{t \to T_{\text{cont}}} \partial_t^k y(t)$ exists strongly in $H^{m-k-1}(\Omega; \mathbb{R}^d)$ and weakly in $H^{m-k}(\Omega; \mathbb{R}^d)$ for $0 \le k \le m - 1$. Since $H^{m-1}(\Omega; \mathbb{R}^d) \hookrightarrow C^1(\overline{\Omega}; \mathbb{R}^d)$ we also have the convergence $y(t) \to \tilde{h}_0$ in $C^1(\overline{\Omega}; \mathbb{R}^d)$. In particular,

$$\overline{\{\nabla \tilde{h}_0(x) \colon x \in \Omega\}} \subset V_{\text{LH}}.$$

Now we want to use the local existence result, Theorem 5.3, with shifted $f, g$ and initial conditions $\tilde{h}_0, \tilde{h}_1$. All we have to do, is to check that the compatibility conditions of order $m$ are satisfied. For $k = 0$ or $k = 1$, we clearly have

$$u_k = \tilde{h}_k - \partial_t^k g(\cdot, T_{\text{cont}}) = \lim_{t \to T_{\text{cont}}} \partial_t^k (y(\cdot, t) - g(\cdot, t)) \in H_0^1.$$

For $2 \leq k \leq m - 1$, we know that $\partial_t^k (y - g)(t)$ converges to $\tilde{h}_k - \partial_t^k g(\cdot, T_{\text{cont}})$ strongly in $H^{m-k-1}(\Omega; \mathbb{R}^d)$ and weakly in $H^{m-k}(\Omega; \mathbb{R}^d)$. Therefore, $\tilde{h}_k - \partial_t^k g(\cdot, T_{\text{cont}}) \in H_0^1(\Omega; \mathbb{R}^d)$. Now we just have to argue inductively that $u_k = \tilde{h}_k - \partial_t^k g(\cdot, T_{\text{cont}})$. If this is already true for all $l < k$, we know in particular that $\partial_t^l (y - g)(t) \rightharpoonup u_l$ in $H^{m-l}(\Omega; \mathbb{R}^d)$. Expressing $\partial_t^k (y - g)(t)$ with the equation in terms of $\partial_t^l (y - g)$, $0 \leq l \leq k - 2$ as in Proposition 5.2, we can thus conclude that $\partial_t^k (y - g)(t) \to u_k$ at least in some weaker sense, e.g. in $H^{m-k-1}(\Omega; \mathbb{R}^d)$. To see this one needs to combine the arguments in Proposition 5.1 with the statement on weak-to-strong continuity in Lemma B.1 with $M = m-1$, $N = k-1$, $L = m-k-1$. Therefore, $f(\cdot, T_{\text{cont}} + \cdot)$, $g(\cdot, T_{\text{cont}} + \cdot)$, $\tilde{h}_0$, and $\tilde{h}_1$ satisfy the compatibility conditions of order $m$.

Hence, we can use Theorem 5.3 to find a $\delta > 0$ and an extension of $y$ to $[0, T_{\text{cont}} + \delta]$, such that

$$y \in \bigcap_{k=0}^{m} C^k\big([T_{\text{cont}}, T_{\text{cont}} + \delta]; H^{m-k}(\Omega; \mathbb{R}^d)\big),$$

$y$ is a solution of the equation on $(T_{\text{cont}}, T_{\text{cont}} + \delta)$ with $y(T_{\text{cont}}) = \tilde{h}_0$ and $\dot{y}(T_{\text{cont}}) = \tilde{h}_1$. Here, $\dot{y}(T_{\text{cont}})$ is to be understood in terms of the values on $[T_{\text{cont}}, T_{\text{cont}} + \delta]$ alone. Furthermore, we have

$$\overline{\{\nabla y(x, t) \colon x \in \Omega, t \in [T_{\text{cont}}, T_{\text{cont}} + \delta]\}} \subset V_{\text{LH}}.$$

We have to take a closer look at what happens in $T_{\text{cont}}$. We clearly have

$$u_k = \lim_{t \to T_{\text{cont}}^+} \partial_t^k (y(\cdot, t) - g(\cdot, t))$$

strongly in $H^{m-k}(\Omega; \mathbb{R}^d)$. But we already saw that $u_k = \tilde{h}_k - \partial_t^k g(\cdot, T_{\text{cont}})$ for $0 \leq k \leq m-1$. So the weak derivatives are continuous, which directly implies the strong differentiability

$$y \in \bigcap_{k=0}^{m-1} C^k\big([0, T_{\text{cont}} + \delta]; H^{m-k-1}(\Omega; \mathbb{R}^d)\big).$$

Furthermore, we have one more strong derivative outside of $T_{\text{cont}}$ which extends to the entire interval including $T_{\text{cont}}$ as a weak derivative. By continuity it is bounded on $[T_{\text{cont}}, T_{\text{cont}} + \delta]$ and we have already shown the boundedness on $[0, T_{\text{cont}})$. Therefore,

$$y \in \bigcap_{k=0}^{m} W^{k,\infty}\big([0, T_{\text{cont}} + \delta]; H^{m-k}(\Omega; \mathbb{R}^d)\big).$$

Additionally, by compactness and identification $\partial_t^k y$ is continuous in $H^{m-k}(\Omega; \mathbb{R}^d)$ with respect to the weak topology for all $0 \leq k \leq m - 1$.

Now, we want to use the ideas of [Str66] to get the missing additional regularity. The key is to use that $y$ solves an equation.

Clearly $v := \partial_t^{m-1}(y - g)$ satisfies $v \in L^\infty(0, T_{\text{cont}} + \delta; H_0^1(\Omega; \mathbb{R}^d))$ with weak derivative $\partial_t v \in L^\infty(0, T_{\text{cont}} + \delta; L^2(\Omega; \mathbb{R}^d))$. We claim that it also has a weak second derivative in $L^\infty(0, T_{\text{cont}} + \delta; H^{-1}(\Omega; \mathbb{R}^d))$. To that end, we calculate

$$
\begin{aligned}
\partial_t^{m-1}&(DW_{\text{CB}}(\nabla y)_{ij}) \\
&= D^2 W_{\text{CB}}(\nabla y)[\nabla v, E_{ij}] + D^2 W_{\text{CB}}(\nabla y)[\partial_t^{m-1}\nabla g, E_{ij}] \\
&\quad + \sum_{\substack{\beta \in \mathbb{N}_0^{d\times d} \\ 2 \leq |\beta| \leq m-1}} \sum_{s=1}^{m-1} \sum_{p_s(m-1,\beta)} (m-1)! D^{\beta + E_{ij}} W_{\text{CB}}(\nabla y) \prod_{j=1}^{s} \frac{(\partial_t^{\gamma_j} \nabla y)^{\lambda_j}}{\lambda_j! \gamma_j!^{|\lambda_j|}} \\
&=: D^2 W_{\text{CB}}(\nabla y)[\nabla v, E_{ij}] + R_{ij}.
\end{aligned}
$$

We can now use Lemma B.1 with $M = m - 2$ and $N = \sum |\lambda_j|(\gamma_j - 1) = m - 1 - |\beta| \leq m - 3$ to see that

$$\prod_{j=1}^{s} (\partial_t^{\gamma_j} \nabla y)^{\lambda_j} \in L^\infty(0, T_{\text{cont}} + \delta; H^1(\Omega; \mathbb{R}^d)).$$

Since

$$D^{\beta + E_{ij}} W_{\text{CB}}(\nabla y) \in L^\infty(0, T_{\text{cont}} + \delta; W^{1,\infty}(\Omega; \mathbb{R}^d)),$$

we obtain

$$R \in L^\infty(0, T_{\text{cont}} + \delta; H^1(\Omega; \mathbb{R}^{d\times d}))$$

and

$$F := \partial_t^{m-1} f - \partial_t^{m+1} g + \operatorname{div} R \in L^\infty(0, T_{\text{cont}} + \delta; L^2(\Omega; \mathbb{R}^d)).$$

Defining $A(t) : H_0^1(\Omega; \mathbb{R}^d) \to H^{-1}(\Omega; \mathbb{R}^d)$ by

$$A(t)u = -\operatorname{div}(D^2 W_{\mathrm{CB}}(\nabla y(\cdot, t))[\nabla u]),$$

we can use a weak formulation (in time and space) of the equation to see that indeed $\partial_t^2 v$ exists as a weak derivative in $L^\infty(0, T_{\mathrm{cont}} + \delta; H^{-1}(\Omega; \mathbb{R}^d))$ and satisfies

$$\partial_t^2 v(t) + A(t)v(t) = F(t).$$

Let us look more precisely at $A$. Since $\nabla y \in C([0, T_{\mathrm{cont}} + \delta] \times \overline{\Omega}; \mathbb{R}^{d \times d})$, the coefficients $D^2 W_{\mathrm{CB}}(\nabla y(x, t))$ are uniformly bounded, uniformly continuous and have a positive, uniform Legendre-Hadamard constant. Under these conditions it is well known that $A(t)$ satisfies a Gårding-inequality uniformly in time, see Theorem C.1. I.e., there are $\lambda_1 > 0$, $\lambda_2 \in \mathbb{R}$ such that

$$\langle A(t)v, v \rangle_{H^{-1}, H_0^1} \geq \lambda_1 \|v\|_{H_0^1} - \lambda_2 \|v\|_{L^2}$$

for all $t$ and all $v \in H_0^1(\Omega; \mathbb{R}^d)$. Given fixed $v_1, v_2 \in H_0^1(\Omega; \mathbb{R}^d)$, $\langle A(t)v_1, v_2 \rangle_{H^{-1}, H_0^1}$ has the weak derivative $\langle A'(t)v_1, v_2 \rangle_{H^{-1}, H_0^1}$, where

$$A'(t)u = -\operatorname{div}(D^3 W(\nabla y(\cdot, t))[\partial_t \nabla y(\cdot, t), \nabla u]).$$

Since $D^3 W(\nabla y)[\partial_t \nabla y] \in L^\infty([0, T_{\mathrm{cont}} + \delta] \times \Omega; \mathbb{R}^{d \times d \times d \times d})$, we see that $A'$ is bounded with values in $L(H_0^1(\Omega; \mathbb{R}^d), H^{-1}(\Omega; \mathbb{R}^d))$. Therefore, we can use Theorem A.2 to conclude that

$$\partial_t^m y \in C([0, T_{\mathrm{cont}} + \delta]; L^2(\Omega; \mathbb{R}^d))$$

and

$$\partial_t^{m-1} y \in C([0, T_{\mathrm{cont}} + \delta]; H^1(\Omega; \mathbb{R}^d)).$$

For $1 \leq k \leq m - 2$, taking $k$ time derivatives in the equation we find

$$\partial_t^k y = (A(t) + \lambda \operatorname{Id})^{-1}(-\partial_t^{k+2} y + \lambda \partial_t^k y + \partial_t^k f + \operatorname{div} S).$$

Here

$$S = \partial_t^k(DW_{\mathrm{CB}}(\nabla y)) - D^2 W_{\mathrm{CB}}(\nabla y)[\partial_t^k \nabla y]$$

$$= \sum_{\substack{\beta \in \mathbb{N}_0^{d \times d} \\ 2 \leq |\beta| \leq k}} \sum_{s=1}^{k} \sum_{p_s(k, \beta)} k! D^{\beta + E_{ij}} W_{\mathrm{CB}}(\nabla y) \prod_{j=1}^{s} \frac{(\partial_t^{\gamma_j} \nabla y)^{\lambda_j}}{\lambda_j! \gamma_j!^{|\lambda_j|}}$$

and $A(t) + \lambda\,\mathrm{Id}\colon H^{m-k}\cap H_0^1 \to H^{m-k-2}$ is invertible for $\lambda$ large enough because of Theorem C.3, where we use that $D^2 W_{\mathrm{CB}}(\nabla y) \in L^\infty([0, T_{\mathrm{cont}}+\delta], H^{m-1}(\Omega; \mathbb{R}^d))$ according to Lemma B.5. Theorem C.3 also gives a time independent bound on $\|(A(t) + \lambda\,\mathrm{Id})^{-1}\|_{L(H^{m-k-2}, H^{m-k})}$.

According to Lemma B.5, $B \mapsto D^3 W_{\mathrm{CB}} \circ B$ is a bounded map from $H^{m-2}$ to $H^{m-2}$. Therefore, we can use Lemma B.1 with $M = m-2$ to see that $A'(t)\colon H^{m-k} \to H^{m-k-2}$ is well defined with

$$\|A'(t)\|_{L(H^{m-k}, H^{m-k-2})} \le C$$

uniform in $t$. Since

$$
\begin{aligned}
(A(t) + \lambda\,\mathrm{Id})^{-1} &- (A(s) + \lambda\,\mathrm{Id})^{-1} \\
&= -(A(t) + \lambda\,\mathrm{Id})^{-1}(A(t) - A(s))(A(s) + \lambda\,\mathrm{Id})^{-1},
\end{aligned}
$$

we obtain

$$\|(A(t) + \lambda\,\mathrm{Id})^{-1} - (A(s) + \lambda\,\mathrm{Id})^{-1}\|_{L(H^{m-k-2}, H^{m-k}\cap H_0^1)} \le C|t - s|.$$

Using that $\partial_t^\gamma \nabla y$ is weakly continuous in $H^{m-1-\gamma}$, we can use Lemma B.1 and its additional statement with $M = m-2$, $N = k - |\beta|$, $L = m - k - 1 < M - N$ and $\lambda_k = \gamma_j - 1$ to find that $S$ is (strongly) continuous with values in $H^{m-k-1}$.

Putting all of this together we find inductively, starting at $k = m$ and $k = m - 1$, that

$$\partial_t^k y \in C([0, T_{\mathrm{cont}} + \delta]; H^{m-k}(\Omega; \mathbb{R}^d))$$

for $1 \le k \le m$.

For $k = 0$ we can no longer use the theory for linear systems in divergence form. Instead, we look at the operator

$$(A(t)u)_i = -\sum_{k,j,l}(D^2 W_{\mathrm{CB}}(\nabla y(\cdot, t)))_{ijkl}\frac{\partial^2 u_k}{\partial x_j \partial x_l}.$$

Now we have

$$y = (A(t) + \lambda\,\mathrm{Id})^{-1}(\partial_t^2 y + \lambda y - f).$$

But Theorem C.3 also holds in non-divergence form, hence

$$(A(t) + \lambda\,\mathrm{Id})^{-1}\colon H^{m-2}(\Omega; \mathbb{R}^d) \to H^m(\Omega; \mathbb{R}^d) \cap H_0^1(\Omega; \mathbb{R}^d)$$

is well defined and bounded independently of $t$ since $m - 2 > \frac{d}{2}$ gives a bound on $D^2 W_{\mathrm{CB}}(\nabla y(\cdot; t))$ in $W^{1,\infty}$. The continuity then follows along the same lines as for $k \geq 1$.

Having established the additional regularity, we have a contradiction to the definition of $T_{\mathrm{cont}}$. This proves the existence of a $T_{\mathrm{cont}}$ with the desired properties. But due to the local uniqueness of solutions, any smaller $T$ cannot satisfy either one of (i), (ii) or (iii). Therefore, $T_{\mathrm{cont}}$ is unique. $\qquad\square$

## 5.2  Atomistic Elastodynamics

The main theorem of this section is the long-time existence of a solution to the atomistic equations (for $\varepsilon$ small enough), whenever the corresponding solution to the continuum equation exists and is atomistically stable. But before we state and prove the theorem, let us prove two auxiliary theorems that are already interesting on their own. In both cases we will give more general versions than what we will actually need for the main theorem.

### A First Local Result

We start with a theorem on local existence and uniqueness.

**Theorem 5.5.** *Let $d \in \{1, 2, 3, 4\}$, $V \subset \mathbb{R}^{d \times \mathcal{R}}$, $W_{\mathrm{atom}} \in C_b^2(V)$ and set $V_{\mathrm{atom}} = \{A \in V : \lambda_{\mathrm{atom}}(A) > 0\}$. Let $\varepsilon_0 > 0$, $C_1 > 0$, $r_0 > 0$ and $\gamma \in \left[\frac{d}{2}, 2\right]$, such that $4 C_1 \varepsilon_0^{\gamma - \frac{d}{2}} \leq r_0$. Furthermore, let $0 < \varepsilon \leq \varepsilon_0$, $T_0 > 0$ and fix $f_{\mathrm{atom}}$, $g_{\mathrm{atom}}$, $y_{\mathrm{ref}}$ with*

$$f_{\mathrm{atom}}(x, \cdot) \in L^2((0, T_0); \mathbb{R}^d) \quad \textit{for all } x \in \mathrm{int}_\varepsilon \, \Omega,$$

$$g_{\mathrm{atom}}(x, \cdot) \in H^2((0, T_0); \mathbb{R}^d) \quad \textit{for all } x \in \partial_\varepsilon \Omega,$$

$$y_{\mathrm{ref}}(x, \cdot) \in H^2((0, T_0); \mathbb{R}^d) \quad \textit{for all } x \in \Omega \cap \varepsilon \mathbb{Z}^d,$$

*such that*

$$\mathrm{dist}(D_{\mathcal{R}, \varepsilon} y_{\mathrm{ref}}(x, t), V_{\mathrm{atom}}^c) > r_0$$

*in* $\mathrm{sint}_\varepsilon \, \Omega \times [0, T_0]$ *and*

$$\sup_t \| y_{\mathrm{ref}}(t) - g_{\mathrm{atom}}(t) \|_{\partial_\varepsilon \Omega} \leq C_1 \varepsilon^\gamma.$$

*Then there exists a time $T > 0$ which may depend on all the previous quantities, including $\varepsilon$, such that the following holds: Given any $t_0 \in [0, T_0)$, $h_{\mathrm{atom},0} \in \mathcal{A}_\varepsilon(\Omega, g_{\mathrm{atom}}(\cdot, t_0))$, and $h_{\mathrm{atom},1} \in \mathcal{A}_\varepsilon(\Omega, \dot{g}_{\mathrm{atom}}(\cdot, t_0))$, such that*

$$\|h_{\mathrm{atom},1} - \dot{y}_{\mathrm{ref}}(\cdot, t_0)\|^2_{\ell^2_\varepsilon(\mathrm{int}_\varepsilon \Omega)} + \|h_{\mathrm{atom},0} - y_{\mathrm{ref}}(\cdot, t_0)\|^2_{h^1_\varepsilon(\mathrm{sint}_\varepsilon \Omega)} \leq C_1^2 \varepsilon^{2\gamma},$$

*there is a unique solution $y \in H^2((t_0, \min\{t_0 + T, T_0\}); \mathbb{R}^d)^{\Omega \cap \varepsilon \mathbb{Z}^d}$ to the discrete initial-boundary-value problem on $[t_0, \min\{t_0 + T, T_0\}]$.*

*Proof.* This is basically the Picard-Lindelöf Theorem. But we want to quantify the dependence on the initial conditions. We look at the set

$$K_{T,b,z_0} = \Big\{ (z_1, z_2) \colon z_1, z_2 \in C([t_0, \min\{t_0 + T, T_0\}]; \ell^\infty(\Omega \cap \varepsilon \mathbb{Z}^d))$$
$$z_1(t) \in \mathcal{A}_\varepsilon(\Omega; 0), z_2(t) \in \mathcal{A}_\varepsilon(\Omega; 0)$$
$$\sup_t \|z(t) - z^0\|_{\ell^\infty(\Omega \cap \varepsilon \mathbb{Z}^d)} \leq b \Big\},$$

with the metric induced by $\|z\| = \sup_t \|z(t)\|_{\ell^\infty(\Omega \cap \varepsilon \mathbb{Z}^d)}$. Here we substituted

$$z(t) = \begin{pmatrix} y(t) - y_{\mathrm{ref}}(t) - T_\varepsilon(g_{\mathrm{atom}}(t) - y_{\mathrm{ref}}(t)) \\ \dot{y}(t) - \dot{y}_{\mathrm{ref}}(t) - T_\varepsilon(\dot{g}_{\mathrm{atom}}(t) - \dot{y}_{\mathrm{ref}}(t)) \end{pmatrix}$$

and

$$z^0 = \begin{pmatrix} h_{\mathrm{atom},0} - y_{\mathrm{ref}}(t_0) - T_\varepsilon(g_{\mathrm{atom}}(t_0) - y_{\mathrm{ref}}(t_0)) \\ h_{\mathrm{atom},1} - \dot{y}_{\mathrm{ref}}(t_0) - T_\varepsilon(\dot{g}_{\mathrm{atom}}(t_0) - \dot{y}_{\mathrm{ref}}(t_0)) \end{pmatrix}.$$

The equation can be written as $\dot{z}(t) = F(t, z(t))$, where $F_1(t, z_1, z_2) = z_2$ and

$$F_2(t, z_1, z_2)(x) = f_{\mathrm{atom}}(x, t) - \ddot{y}_{\mathrm{ref}}(x, t) - T_\varepsilon(\ddot{g}_{\mathrm{atom}}(t) - \ddot{y}_{\mathrm{ref}}(t))$$
$$+ \mathrm{div}_{\mathcal{R},\varepsilon} \left( DW_{\mathrm{atom}}(D_{\mathcal{R},\varepsilon} y_{\mathrm{ref}}(x, t) + D_{\mathcal{R},\varepsilon} T_\varepsilon(g_{\mathrm{atom}}(t) - y_{\mathrm{ref}}(t))(x) \right.$$
$$\left. + D_{\mathcal{R},\varepsilon} z_1(x)) \right)$$

for $x \in \mathrm{int}_\varepsilon \Omega$, but $F_2(t, z_1, z_2)(x) = 0$ for $x \in \partial_\varepsilon \Omega$. Since we do not even claim strong differentiability, it is best to look at the fixed point equation of

$$G(z)(t) = z^0 + \int_{t_0}^t F(s, z(s)) \, ds.$$

Clearly,

$$\sup_t \|y_{\mathrm{ref}}(t) - g_{\mathrm{atom}}(t)\|_{\partial_\varepsilon \Omega} \leq C_1 \varepsilon^\gamma$$

implies

$$\begin{aligned}
|D_{\mathcal{R},\varepsilon}T_\varepsilon(g_{\text{atom}} - y_{\text{ref}})(x,t)| &\leq \varepsilon^{-\frac{d}{2}}\|D_{\mathcal{R},\varepsilon}T_\varepsilon(g_{\text{atom}} - y_{\text{ref}})(t)\|_{h_\varepsilon^1} \\
&= \varepsilon^{-\frac{d}{2}}\|g_{\text{atom}} - y_{\text{ref}}(t)\|_{\partial_\varepsilon\Omega} \\
&\leq C_1\varepsilon^{\gamma-\frac{d}{2}}
\end{aligned}$$

uniformly in $x$ and $t$. Now, if $0 < b \leq \frac{\varepsilon r_0}{8|\mathcal{R}|^{\frac{1}{2}}}$ then for any $z \in K_{T,b,z_0}$

$$\begin{aligned}
&|D_{\mathcal{R},\varepsilon}T_\varepsilon(g_{\text{atom}} - y_{\text{ref}})(x,t) + D_{\mathcal{R},\varepsilon}z_1(x,t)| \\
&\leq |D_{\mathcal{R},\varepsilon}T_\varepsilon(g_{\text{atom}} - y_{\text{ref}})(x,t)| + |D_{\mathcal{R},\varepsilon}(z_1(x,t) - z_1^0(x))| + |D_{\mathcal{R},\varepsilon}z_1^0(x)| \\
&\leq C_1\varepsilon^{\gamma-\frac{d}{2}} + \frac{2b|\mathcal{R}|^{\frac{1}{2}}}{\varepsilon} + |D_{\mathcal{R},\varepsilon}z_1^0(x)| \\
&\leq 3C_1\varepsilon^{\gamma-\frac{d}{2}} + \frac{2b|\mathcal{R}|^{\frac{1}{2}}}{\varepsilon} \\
&\leq r_0.
\end{aligned}$$

Therefore $F(s, z(s))$ is well defined. Furthermore,

$$\begin{aligned}
&\sup_t \sup_{x\in\Omega\cap\varepsilon\mathbb{Z}^d} |G(z)(x,t) - z^0(x)| \\
&\leq \sup_{t,x} \int_{t_0}^t |F_1(s, z(s))| + |F_2(s, z(s))| \, ds \\
&\leq bT + TC_1\varepsilon^{\gamma-\frac{d}{2}} + T\|T_\varepsilon(\dot{g}_{\text{atom}} - \dot{y}_{\text{ref}})\|_{L^\infty(0,T_0;\ell^\infty)} \\
&\quad + T\frac{2|\mathcal{R}|}{\varepsilon}\|DW_{\text{atom}}\|_\infty + \sqrt{T}\|f\|_{L^2(0,T_0;\ell^\infty)} + \sqrt{T}\|\ddot{y}_{\text{ref}}\|_{L^2(0,T_0;\ell^\infty)} \\
&\quad + \sqrt{T}\|T_\varepsilon(\ddot{g}_{\text{atom}} - \ddot{y}_{\text{ref}})\|_{L^2(0,T_0;\ell^\infty)}.
\end{aligned}$$

In particular, for $T$ small enough

$$\sup_t \sup_{x\in\Omega\cap\varepsilon\mathbb{Z}^d} |G(z)(x,t) - z^0(x)| \leq b.$$

Since $G(z)$ also has the correct boundary values, $G\colon K_{T,b,z_0} \to K_{T,b,z_0}$ is well defined. Given $z, \tilde{z} \in K_{T,b,z_0}$ we calculate

$$
\sup_t \sup_{x \in \Omega \cap \varepsilon \mathbb{Z}^d} |G(z)(x,t) - G(\tilde{z})(x,t)| \leq T \sup_t \|F(t, z(t)) - F(t, \tilde{z}(t))\|_{\ell^\infty}
$$

$$
\leq T \Big( \sup_t \|z_2(t) - \tilde{z}_2(t)\|_{\ell^\infty}
$$

$$
+ |\mathcal{R}|^{\frac{3}{2}} \frac{4}{\varepsilon^2} \|D^2 W_{\text{atom}}\|_\infty \sup_t \|z_1(t) - \tilde{z}_1(t)\|_{\ell^\infty} \Big)
$$

$$
\leq T \big( 1 + |\mathcal{R}|^{\frac{3}{2}} \frac{4}{\varepsilon^2} \|D^2 W_{\text{atom}}\|_\infty \big) \sup_t \|z(t) - \tilde{z}(t)\|_{\ell^\infty}
$$

$$
\leq \frac{1}{2} \sup_t \|z(t) - \tilde{z}(t)\|_{\ell^\infty(\Omega \cap \varepsilon \mathbb{Z}^d)},
$$

if we also require

$$
T \leq \frac{1}{2 + 2|\mathcal{R}|^{\frac{3}{2}} \frac{4}{\varepsilon^2} \|D^2 W_{\text{atom}}\|_\infty}.
$$

Now we can use the Banach fixed point theorem. If $b$ and $T$ satisfy the constraints above, then $G$ has a unique fixed point $z \in K_{T,b,z_0}$. Setting $y = z_1 + y_{\text{ref}} + T_\varepsilon(g_{\text{atom}}(t) - y_{\text{ref}}(t))$, we have

$$
y \in H^2((t_0, \min\{t_0 + T, T_0\}); \ell^\infty(\Omega \cap \varepsilon \mathbb{Z}^d))
$$

and $y$ solves the discrete initial-boundary-value problem in the absolutely continuous sense on $[t_0, \min\{t_0 + T, T_0\}]$. Now conversely, if $y$ is any solution in $H^2((t_0, \min\{t_0 + T, T_0\}); \ell^\infty(\Omega \cap \varepsilon \mathbb{Z}^d))$ that satisfies

$$
D_{\mathcal{R},\varepsilon} y(x,t) \in V
$$

for all $t$ and $x \in \text{sint}_\varepsilon \Omega$, we can substitute back to $z$ and calculate

$$
\|z(t) - z^0\|_{\ell^\infty(\Omega \cap \varepsilon \mathbb{Z}^d)}
$$

$$
\leq \int_{t_0}^t \|z(s) - z^0\|_{\ell^\infty(\Omega \cap \varepsilon \mathbb{Z}^d)} \, ds + \frac{2\mathcal{R} \|DW_{\text{atom}}\|_\infty}{\varepsilon} (t - t_0)
$$

$$
+ \sqrt{t - t_0} \big( \|f\|_{L^2(0,T_0;\ell^\infty)} + \|\ddot{y}_{\text{ref}}\|_{L^2(0,T_0;\ell^\infty)}
$$

$$
+ \big\| T_\varepsilon(\ddot{g}_{\text{atom}} - \ddot{y}_{\text{ref}}) \big\|_{L^2(0,T_0;\ell^\infty)} \big)
$$

$$
\leq \int_{t_0}^t \|z(s) - z^0\|_{\ell^\infty(\Omega \cap \varepsilon \mathbb{Z}^d)} \, ds + C_1 \big( (t - t_0) + \sqrt{t - t_0} \big)
$$

Using Grönwall's inequality we thus get

$$\|z(t) - z^0\|_{\ell^\infty(\Omega \cap \varepsilon \mathbb{Z}^d)} \leq C_1(t - t_0 + \sqrt{t - t_0})e^{t-t_0}$$
$$\leq 2C_1\sqrt{T}e^T$$
$$\leq b$$

if we additionally assume $T \leq 1$ and $T \leq \frac{b}{2C_1 e}$. Therefore, $z \in K_{T,b,z_0}$, and the uniqueness of the solution follows. $\qquad\square$

Although this lemma already gives us a local solution, the time $T$ depends heavily on $\varepsilon$ and is not necessarily bounded from below as $\varepsilon$ goes to 0. One of our main goals is to show existence on an $\varepsilon$-independent time interval. Actually, as discussed in the introduction, we even want to go one step further. We will show that the atomistic solution exists as long as the solution to the continuous problem exists and is atomistically stable.

## The Atomistic Gårding Inequality

As argued in the introduction establishing an atomistic Gårding inequality is key to provide control of the stability of solutions for long times and large deformations. There are some clear differences to the continuous Gårding inequality (Theorem C.1) that already make the precise formulation of the theorem challenging. By now unsurprisingly, we need to require atomistic stability. Additionally, due to the discreteness of the problem, we can not just impose continuity of the coefficients which is one of the crucial assumptions in the continuous case. Instead, we need to track the variation of the coefficients and the dependence on $\varepsilon$ more explicitly.

**Theorem 5.6.** *Let $d \in \mathbb{N}$, $\Omega \subset \mathbb{R}^d$ open and bounded, and $\lambda_1, \Lambda, \varepsilon_0 > 0$. Consider a family $A_\varepsilon$: $\mathrm{sint}_\varepsilon \Omega \to \mathbb{R}^{d \times \mathcal{R} \times d \times \mathcal{R}}$, for $0 < \varepsilon \leq \varepsilon_0$, with $\lambda_{\mathrm{atom}}(A_\varepsilon(x)) \geq \lambda_1$ for all $x \in \mathrm{sint}_\varepsilon \Omega$ and $0 < \varepsilon \leq \varepsilon_0$. Assume also that $\sup_\varepsilon \|A_\varepsilon\|_\infty \leq \Lambda$ and that there are $r_\varepsilon \geq \varepsilon$ such that*

$$\sup_{\substack{0 < \varepsilon \leq \varepsilon_0}} \sup_{\substack{x,x' \in \mathrm{sint}_\varepsilon \Omega \\ |x-x'| \leq 2r_\varepsilon + 2\varepsilon R_{\max}}} |A_\varepsilon(x) - A_\varepsilon(x')| \leq \frac{\lambda_1}{4},$$

*then there is a $\lambda_2 = \lambda_2(\lambda_1, \Lambda, d, \mathcal{R})$, such that*

$$\varepsilon^d \sum_{x \in \mathrm{sint}_\varepsilon \Omega} A_\varepsilon(x)[D_{\mathcal{R},\varepsilon}u(x), D_{\mathcal{R},\varepsilon}u(x)] \geq \frac{\lambda_1}{2}\|u\|_{h^1_\varepsilon(\mathrm{sint}_\varepsilon \Omega)}^2 - \frac{\lambda_2}{r_\varepsilon^2}\|u\|_{\ell^2_\varepsilon(\mathrm{int}_\varepsilon \Omega)}^2$$

*for all* $u \in \mathcal{A}_\varepsilon(\Omega, 0)$ *and* $0 < \varepsilon \leq \varepsilon_0$.

*Remark* 5.7. In this thesis we will only use the theorem in the case where $r_\varepsilon$ is independent of $\varepsilon$. This corresponds to $A_\varepsilon$ only changing on the macroscopic scale. We will still prove the more general version since the theorem has some interest itself.

*Proof.* By the definition of atomistic stability we have

$$\varepsilon^d \sum_{x \in \operatorname{sint}_\varepsilon \Omega} A_\varepsilon(z)[D_{\mathcal{R},\varepsilon} u(x), D_{\mathcal{R},\varepsilon} u(x)] \geq \lambda_1 \|u\|^2_{h^1_\varepsilon(\operatorname{sint}_\varepsilon \Omega)}$$

for every $z \in \operatorname{sint}_\varepsilon \Omega$, every $\varepsilon > 0$, and every $u \in \mathcal{A}_\varepsilon(\Omega, 0)$.

Now, choose countable many $z_j \in \mathbb{R}^d$ and a partition of unity $\eta_j \in C_c^\infty(\mathbb{R}^d; [0,1])$ such that $\sum_j \eta_j^2(x) = 1$ for every $x \in \mathbb{R}^d$, $\operatorname{supp} \eta_j \subset B_{r_\varepsilon}(z_j)$, $|\nabla \eta_j| \leq \frac{C(d)}{r_\varepsilon}$, and the decomposition is locally finite in the sense that

$$|\{j \colon B_{r_\varepsilon}(z_j) \cap B_R(x) \neq \emptyset\}| \leq C(d)\left(1 + \frac{R}{r_\varepsilon}\right)^d$$

for all $x \in \mathbb{R}^d$ and $R > 0$. Whenever $B_{r_\varepsilon + \varepsilon R_{\max}}(z_j) \cap \operatorname{sint}_\varepsilon \Omega \neq \emptyset$ fix a point $x_{j,\varepsilon} \in B_{r_\varepsilon + \varepsilon R_{\max}}(z_j) \cap \operatorname{sint}_\varepsilon \Omega$. By assumption we then have $|A_\varepsilon(x_{j,\varepsilon}) - A_\varepsilon(x)| \leq \frac{\lambda_1}{4}$ for every $x \in B_{r_\varepsilon + \varepsilon R_{\max}}(z_j) \cap \operatorname{sint}_\varepsilon \Omega$. Now, since

$$(D_{\mathcal{R},\varepsilon}(\eta_j u)(x))_\rho = \eta_j(x)(D_{\mathcal{R},\varepsilon} u(x))_\rho + u(x + \varepsilon\rho)(D_{\mathcal{R},\varepsilon}\eta_j(x))_\rho$$

for any $\delta > 0$ we can calculate with Young's inequality

$$
\begin{aligned}
\varepsilon^d &\sum_{x \in \text{sint}_\varepsilon \Omega} A_\varepsilon(x)[D_{\mathcal{R},\varepsilon}u(x), D_{\mathcal{R},\varepsilon}u(x)] \\
&= \varepsilon^d \sum_j \sum_{x \in \text{sint}_\varepsilon \Omega} A_\varepsilon(x)[\eta_j(x)D_{\mathcal{R},\varepsilon}u(x), \eta_j(x)D_{\mathcal{R},\varepsilon}u(x)] \\
&\geq \varepsilon^d \sum_j \sum_{x \in \text{sint}_\varepsilon \Omega} A_\varepsilon(x)[D_{\mathcal{R},\varepsilon}(\eta_j u)(x), D_{\mathcal{R},\varepsilon}(\eta_j u)(x)] \\
&\quad - \delta\varepsilon^d \sum_j \sum_{x \in \text{sint}_\varepsilon \Omega} n_j^2(x)|D_{\mathcal{R},\varepsilon}u(x)|^2 \\
&\quad - \Lambda(1 + \frac{\Lambda}{\delta})\varepsilon^d \sum_j \sum_{x \in \text{sint}_\varepsilon \Omega} \sum_\rho |u(x+\varepsilon\rho)|^2 \left|\frac{\eta_j(x+\varepsilon\rho) - \eta_j(x)}{\varepsilon}\right|^2 \\
&\geq \varepsilon^d \sum_j \sum_{x \in \text{sint}_\varepsilon \Omega} A_\varepsilon(x_{j,\varepsilon})[D_{\mathcal{R},\varepsilon}(\eta_j u)(x), D_{\mathcal{R},\varepsilon}(\eta_j u)(x)] \\
&\quad - \frac{\lambda_1}{4}|D_{\mathcal{R},\varepsilon}(\eta_j u)(x)|^2 - \delta\varepsilon^d \sum_{x \in \text{sint}_\varepsilon \Omega} |D_{\mathcal{R},\varepsilon}u(x)|^2 \\
&\quad - \Lambda(1 + \frac{\Lambda}{\delta})\varepsilon^d \sum_j \sum_{x \in \text{sint}_\varepsilon \Omega} \sum_\rho |u(x+\varepsilon\rho)|^2 \left|\frac{\eta_j(x+\varepsilon\rho) - \eta_j(x)}{\varepsilon}\right|^2.
\end{aligned}
$$

But we have the estimate for constant coefficients $A_\varepsilon(x_{j,\varepsilon})$. Therefore,

$$
\begin{aligned}
\varepsilon^d &\sum_{x \in \text{sint}_\varepsilon \Omega} A_\varepsilon(x)[D_{\mathcal{R},\varepsilon}u(x), D_{\mathcal{R},\varepsilon}u(x)] \\
&\geq \varepsilon^d \frac{3}{4}\lambda_1 \sum_j \sum_{x \in \text{sint}_\varepsilon \Omega} |D_{\mathcal{R},\varepsilon}(\eta_j u)(x)|^2 - \delta\varepsilon^d \sum_{x \in \text{sint}_\varepsilon \Omega} |D_{\mathcal{R},\varepsilon}u(x)|^2 \\
&\quad - \Lambda(1 + \frac{\Lambda}{\delta})\varepsilon^d \sum_j \sum_{x \in \text{sint}_\varepsilon \Omega} \sum_\rho |u(x+\varepsilon\rho)|^2 \left|\frac{\eta_j(x+\varepsilon\rho) - \eta_j(x)}{\varepsilon}\right|^2 \\
&\geq \varepsilon^d (\frac{3}{4}\lambda_1 - 2\delta) \sum_{x \in \text{sint}_\varepsilon \Omega} |D_{\mathcal{R},\varepsilon}u(x)|^2 - \left(\Lambda(1 + \frac{\Lambda}{\delta}) + \frac{3}{4}\lambda_1(1 + \frac{3\lambda_1}{4\delta})\right) \\
&\quad \cdot \|u\|_{\ell_\varepsilon^2(\text{int}_\varepsilon \Omega)}^2 C(d, \mathcal{R}) \frac{1}{r_\varepsilon^2}\left(1 + \frac{\varepsilon R_{\max}}{r_\varepsilon}\right)^d
\end{aligned}
$$

Now, choosing $\delta = \frac{\lambda_1}{8}$ and using $r_\varepsilon \geq \varepsilon$, we indeed get

$$\varepsilon^d \sum_{x \in \text{sint}_\varepsilon \Omega} A_\varepsilon(x)[D_{\mathcal{R},\varepsilon}u(x), D_{\mathcal{R},\varepsilon}u(x)]$$

$$\geq \frac{\lambda_1}{2}\varepsilon^d \sum_{x \in \text{sint}_\varepsilon \Omega} |D_{\mathcal{R},\varepsilon}u(x)|^2 - C(\lambda_1, \Lambda, d, \mathcal{R})\frac{1}{r_\varepsilon^2}\|u\|_{\ell_\varepsilon^2(\text{int}_\varepsilon \Omega)}^2.$$

$\square$

## The Main Theorem on Atomistic Elastodynamics

Let us make some last preparations for our main theorem. We will show that there are atomistic solutions close to the extended and regularized reference configuration

$$y_{\text{ref}} = \eta_\varepsilon * (E y_{\text{cont}})$$

where $y_{\text{cont}}$ is a solution of the continuous problem, $\eta_\varepsilon(x)$ denotes the standard scaled mollifying kernel, and $E$ denotes the Stein extension which is an extension operator for all Sobolev spaces requiring only very little regularity of the boundary, cf. [Ste70, Chapter VI].

The conditions that we will pose on the time-dependent atomistic boundary conditions can be formulated much easier with the following norm. Given $g \colon \partial_\varepsilon \Omega \times [0, T_0]$, such that $g(x, \cdot) \in H^2(0, T_0)$ for all $x \in \partial_\varepsilon\Omega$, we look at the (quadratic) functional

$$\mathcal{F}(z) = \|z(0)\|_{\ell_\varepsilon^2(\text{int}_\varepsilon \Omega)}^2 + \|z(0)\|_{h_\varepsilon^1(\text{sint}_\varepsilon \Omega)}^2 + \|\dot{z}(0)\|_{\ell_\varepsilon^2(\text{int}_\varepsilon \Omega)}^2$$

$$+ \int_0^{T_0} \|z(\tau)\|_{h_\varepsilon^1(\text{sint}_\varepsilon \Omega)}^2 + \|\dot{z}(\tau)\|_{h_\varepsilon^1(\text{sint}_\varepsilon \Omega)}^2 + \|\ddot{z}(\tau)\|_{\ell_\varepsilon^2(\text{int}_\varepsilon \Omega)}^2 \, d\tau$$

for $z \colon \Omega \cap \varepsilon\mathbb{Z}^d \times [0, T_0]$, such that $z(x, \cdot) \in H^2(0, T_0)$ for all $x \in \Omega \cap \varepsilon\mathbb{Z}^d$ and $z|_{\partial_\varepsilon\Omega \times [0,T_0]} = g$. Clearly the functional is lower semi-continuous and coercive in $H^2$ and thus has a minimizer. By strict convexity this minimizer is unique and it is also given as the unique solution to

$$0 = (z(0), w(0))_{\ell_\varepsilon^2}^2 + (z(0), w(0))_{h_\varepsilon^1}^2 + (\dot{z}(0), \dot{w}(0))_{\ell_\varepsilon^2}^2$$

$$+ \int_0^{T_0} (z(\tau), w(\tau))_{h_\varepsilon^1}^2 + (\dot{z}(\tau), \dot{w}(\tau))_{h_\varepsilon^1}^2 + (\ddot{z}(\tau), \ddot{w}(\tau))_{\ell_\varepsilon^2}^2 \, d\tau$$

for all $w \in H^2$ with $w|_{\partial_\varepsilon\Omega} = 0$. In particular, the mapping $K_\varepsilon$ that maps $g$ to this minimizer is linear. Furthermore, $\|g\|_{\partial_\varepsilon\Omega, dyn} := \left(\mathcal{F}(K_\varepsilon g)\right)^{\frac{1}{2}}$ is a norm. Besides dominating the norms used in its definition, we will also use that

$$\|K_\varepsilon g\|_{L^\infty(0,T_0;h_\varepsilon^1)} \leq \|g\|_{\partial_\varepsilon\Omega,dyn}$$

and

$$\|K_\varepsilon g\|_{W^{1,\infty}(0,T_0;\ell_\varepsilon^2)} \leq C(T)\|g\|_{\partial_\varepsilon\Omega,dyn}.$$

We will then require

$$\|y_{\text{ref}} - g_{\text{atom}}\|_{\partial_\varepsilon\Omega,dyn} \leq C_g\varepsilon^\gamma,$$

in our main theorem below for some convergence rate $\gamma \in (\frac{d}{2}, 2]$.

While this specific norm is mainly chosen to satisfy certain inequalities in the proof, it is not at all surprising. The terms at the starting time are obviously required by the convergence estimate we want to prove uniformly in time (see below). The terms controlling the $h_\varepsilon^1$-norm are crucial. Among other things, they ensure the uniform convergence of the gradients. Therefore, at the boundary, the atomistic boundary conditions enforce not only the correct asymptotic boundary values but also the correct asymptotic (normal) derivative and thus suppress surface relaxation effects. This is important for the Cauchy-Born rule to hold near the boundary. At last, a difference in the second time derivatives has a similar effect as a difference in the body forces and thus, unsurprisingly, we want both terms to be small in the same norm.

**Theorem 5.8.** *Let $d \in \{2,3\}$ and $m \in \mathbb{N}$, $m \geq 4$. Let $T_0 > 0$ and let $\Omega \subset \mathbb{R}^d$ be an open, bounded set with $\partial\Omega$ of class $C^m$. Let $V \subset \mathbb{R}^{d\times\mathcal{R}}$ be open and $W_{\text{atom}} \in C_b^{m+1}(V)$. Let $f$ be a continuous body force, $h_0, h_1$ initial data and $g$ boundary values such that*

$$f \in C^{m-1}(\overline{\Omega} \times [0,T_0]; \mathbb{R}^d)$$
$$g \in C^{m+1}(\overline{\Omega} \times [0,T_0]; \mathbb{R}^d)$$
$$h_0 \in H^m(\Omega; \mathbb{R}^d)$$
$$\{(\nabla h_0(x)\rho)_{\rho\in\mathcal{R}} \colon x \in \overline{\Omega}\} \subset V \cap \{A \colon \lambda_{\text{atom}}(A) > 0\}$$
$$h_1 \in H^{m-1}(\Omega; \mathbb{R}^d)$$

*and such that the compatibility conditions of order $m$ are satisfied. Furthermore, assume that the unique solution of the Cauchy-Born problem $y_{\text{cont}}$ from Theorem 5.3 exists until $T_0$ and satisfies*

$$y_{\text{cont}} \in \bigcap_{k=0}^{m} C^k\big([0,T_0]; H^{m-k}(\Omega; \mathbb{R}^d)\big),$$

*as well as*

$$\{(\nabla y_{\mathrm{cont}}(x,t)\rho)_{\rho \in \mathcal{R}} \colon x \in \overline{\Omega}, t \in [0,T_0]\} \subset V \cap \{A \colon \lambda_{\mathrm{atom}}(A) > 0\}.$$

*Now let $C_g, C_f, C_h > 0$ and $\gamma \in (\frac{d}{2} + \frac{1}{m-1}, 2]$. Then there is an $\varepsilon_0 > 0$ such that the following holds for every $0 < \varepsilon \leq \varepsilon_0$.*

*Given atomistic body forces $f_{\mathrm{atom}} \in L^2((0,T_0); \mathbb{R}^d)^{\mathrm{int}_\varepsilon \Omega}$, atomistic boundary values $g_{\mathrm{atom}} \in H^2((0,T_0); \mathbb{R}^d)^{\partial_\varepsilon \Omega}$, and atomistic initial data $h_{\mathrm{atom},0} \in \mathcal{A}_\varepsilon(\Omega, g_{\mathrm{atom}}(\cdot,0)), h_{\mathrm{atom},1} \in \mathcal{A}_\varepsilon(\Omega, \dot{g}_{\mathrm{atom}}(\cdot,0))$ with*

$$\|f_{\mathrm{ref}} - f_{\mathrm{atom}}\|_{L^2(0,T_0; \ell_\varepsilon^2(\mathrm{int}\,\Omega))} \leq C_f \varepsilon^\gamma,$$

$$\|y_{\mathrm{ref}} - g_{\mathrm{atom}}\|_{\partial_\varepsilon \Omega, dyn} \leq C_g \varepsilon^\gamma,$$

$$\|h_{\mathrm{atom},1} - \dot{y}_{\mathrm{ref}}(0)\|_{\ell_\varepsilon^2(\mathrm{int}_\varepsilon \Omega)}^2 + \|h_{\mathrm{atom},0} - y_{\mathrm{ref}}(0)\|_{\ell_\varepsilon^2(\mathrm{int}_\varepsilon \Omega)}^2$$
$$+ \|h_{\mathrm{atom},0} - y_{\mathrm{ref}}(0)\|_{h_\varepsilon^1(\mathrm{int}_\varepsilon \Omega)}^2 \leq C_h^2 \varepsilon^{2\gamma},$$

*where*

$$f_{\mathrm{ref}} = \tilde{f} + \ddot{y}_{\mathrm{ref}} - \ddot{y}_{\mathrm{cont}}.$$

*Then there is a unique $y \in H^2((0,T_0); \mathbb{R}^d)^{\Omega \cap \varepsilon \mathbb{Z}^d}$ that solves the atomistic equations with body force $f_{\mathrm{atom}}$ boundary values $g_{\mathrm{atom}}$ and initial conditions $h_{\mathrm{atom},0}, h_{\mathrm{atom},1}$. Furthermore, we have the convergence estimate*

$$\|\dot{y} - \dot{y}_{\mathrm{ref}}\|_{\ell_\varepsilon^2(\mathrm{int}_\varepsilon \Omega)} + \|y - y_{\mathrm{ref}}\|_{h_\varepsilon^1(\mathrm{sint}_\varepsilon \Omega)} + \|y - y_{\mathrm{ref}}\|_{\ell_\varepsilon^2(\mathrm{int}_\varepsilon \Omega)}$$
$$\leq C e^{Ct}(C_g + C_h + C_f + \varepsilon^{2-\gamma})\varepsilon^\gamma$$

*for some $C = C(\mathcal{R}, V, W_{\mathrm{atom}}, y_{\mathrm{cont}}, \Omega, m, \gamma) > 0$.*

*Remark* 5.9. Remember that

$$\tilde{f}(x) = \fint_{Q_\varepsilon(x)} f(z)\, dz.$$

If $y_{\mathrm{cont}} \in H^2(0,T; C^{1,\gamma-1}(\Omega; \mathbb{R}^d))$, the more natural choice $f_{\mathrm{ref}} = \tilde{f}$ suffices since then

$$\|\ddot{y}_{\mathrm{ref}} - \ddot{y}_{\mathrm{cont}}\|_{L^2(0,T_0; \ell_\varepsilon^2(\mathrm{int}\,\Omega))} \leq C(y_{\mathrm{cont}})\varepsilon^\gamma.$$

This condition is automatically satisfied if $m \geq 6$.

*Proof.* First let us prove that $E y_{\mathrm{cont}}$ and $y_{\mathrm{ref}}$ inherit the atomistic stability from $y_{\mathrm{cont}}$ as long as we stay in or close to $\Omega$. Given $R > 0$ and $x \in \Omega + B_{\varepsilon R}(0)$, take $x' \in \Omega$ with $|x - x'| \leq R\varepsilon$. Then we directly see

$$|\nabla E y_{\mathrm{cont}}(x) - \nabla y_{\mathrm{cont}}(x')| \leq \|\nabla^2 E y_{\mathrm{cont}}\|_{L^\infty} R\varepsilon$$
$$\leq C(\Omega) R\varepsilon \|\nabla^2 y_{\mathrm{cont}}\|_{L^\infty}$$

since $y_{\text{cont}} \in H^4(\Omega)$, which embeds into $W^{2,\infty}$ and even $C^2$ for $d \leq 3$. It immediately follows that

$$|\nabla y_{\text{ref}}(x) - \nabla y_{\text{cont}}(x')| \leq C(\Omega)(R+1)\varepsilon\|\nabla^2 y_{\text{cont}}\|_{L^\infty}$$

and

$$\begin{aligned}
&|D_{\mathcal{R},\varepsilon} y_{\text{ref}}(x) - (\nabla y_{\text{cont}}(x')\rho)_{\rho \in \mathcal{R}}| \\
&= \left(\sum_\rho \left|\int_0^1 \nabla y_{\text{ref}}(x + s\varepsilon\rho)\rho - \nabla y_{\text{cont}}(x')\rho\, ds\right|^2\right)^{\frac{1}{2}} \\
&\leq C(\Omega, \mathcal{R}, R)\varepsilon\|\nabla^2 y_{\text{cont}}\|_{L^\infty} \\
&= C(\Omega, \mathcal{R}, R, y_{\text{cont}})\varepsilon
\end{aligned}$$

Since the stability constant is continuous, the set $\{A \in V : \lambda_{\text{atom}}(A) > 0\}$ is open. On the other hand, $\{(\nabla y_{\text{cont}}(x,t)\rho)_{\rho \in \mathcal{R}} : x \in \overline{\Omega}, t \in [0,T]\}$ is compact. Therefore,

$$\begin{aligned}
\{(\nabla y_{\text{cont}}(x,t)\rho)_{\rho \in \mathcal{R}} : x \in \overline{\Omega}, t \in [0,T]\} &+ \overline{B_{\varepsilon C}(0)} \\
&\subset \{A \in V : \lambda_{\text{atom}}(A) > 0\}
\end{aligned}$$

for all $\varepsilon \leq \varepsilon_0$ if $\varepsilon_0 = \varepsilon_0(\mathcal{R}, \Omega, y_{\text{cont}}, R, V, W_{\text{atom}})$ is chosen small enough.

For a time dependent atomistic deformation we define the norm-energy

$$\mathcal{E}(t) = \|\dot{u}\|^2_{\ell^2_\varepsilon(\text{int}_\varepsilon \Omega)} + \|u\|^2_{h^1_\varepsilon(\text{sint}_\varepsilon \Omega)} + \|u\|^2_{\ell^2_\varepsilon(\text{int}_\varepsilon \Omega)},$$

where $u = y - y_{\text{ref}} - K_\varepsilon(g_{\text{atom}} - y_{\text{ref}})$. Note that this energy is well-defined and continuous on $[a,b]$ if $u \in H^2((a,b); \mathbb{R}^d)^{\Omega \cap \varepsilon \mathbb{Z}^d}$.

For $B > 0$ to be defined later, let $T_\varepsilon$ be the supremum of all times $T \leq T_0$ such that a solution $y$ exists on $[0,T)$ and

$$\mathcal{E}(t) \leq B^2 \varepsilon^{2\gamma}$$

for $t \in [0,T]$.

Note that

$$\sup_t \|y_{\text{ref}}(t) - g_{\text{atom}}(t)\|_{\partial_\varepsilon \Omega} \leq \|y_{\text{ref}} - g_{\text{atom}}\|_{\partial_\varepsilon \Omega, dyn} \leq C_g \varepsilon^\gamma.$$

Choosing $\varepsilon_0$ so small that

$$4(\max\{C_g, C_h\} + 1)\varepsilon_0^{\gamma - \frac{d}{2}} \leq \inf_{x,t} \text{dist}(D_{\mathcal{R},\varepsilon} y_{\text{ref}}(x,t), V^c_{\text{atom}})$$

we can apply the local result, Theorem 5.5. If $B > \sqrt{2C_g^2 + 2C_h^2}$, which
will be the case in our choice of $B$, then we indeed see that $T_\varepsilon > 0$. The
uniqueness part of Theorem 5.5 implies that all such solutions agree on
the intersection of their domains of definition. Putting these solutions
together we thus have a $y$ on $(0, T_\varepsilon)$ such that for every $0 < T < T_\varepsilon$ it
holds that $y \in H^2(0, T)$ and $y$ is a solution of the problem. If we choose
$\varepsilon_0$ even smaller, such that

$$4(\sqrt{2B^2 + 2C_g^2} + 1)\varepsilon_0^{\gamma - \frac{d}{2}} \le \inf_{x,t} \operatorname{dist}(D_{\mathcal{R},\varepsilon}y_{\mathrm{ref}}(x,t), V_{\mathrm{atom}}^c)$$

we can again apply Theorem 5.5 with $t_0 \in (0, T_\varepsilon)$ and initial conditions
$y(t_0), \dot{y}(t_0)$, since

$$\|\dot{y}(t_0) - \dot{y}_{\mathrm{ref}}(t_0)\|_{\ell_\varepsilon^2(\mathrm{int}_\varepsilon \, \Omega)}^2 + \|y(t_0) - y_{\mathrm{ref}}(t_0)\|_{h_\varepsilon^1(\mathrm{sint}_\varepsilon \, \Omega)}^2 \le 2\mathcal{E}(t) + 2C_g^2 \varepsilon^{2\gamma}$$
$$\le (2B^2 + 2C_g^2)\varepsilon^{2\gamma}.$$

We thus get a solution on $(t_0, \max\{t_0 + T_{\mathrm{loc}}, T_0\})$ for some $T_{\mathrm{loc}}$ independent of $t_0$. Again by uniqueness all solutions fit together. Therefore,
$y \in H^2(0, T_\varepsilon)$ and $y$ is a solution of the problem on $(0, T_\varepsilon)$. Additionally,
we know that $T_\varepsilon = T_0$ or the solution exists on a larger interval than
$(0, T_\varepsilon)$. In the second case we must have $\mathcal{E}(T_\varepsilon) = B^2\varepsilon^{2\gamma}$. To ensure that
we are in the first case it thus suffices to estimate the energy on $[0, T_\varepsilon]$.
This is what we will do in the rest of the proof.

The energy bound implies

$$\|D_{\mathcal{R},\varepsilon}y - D_{\mathcal{R},\varepsilon}y_{\mathrm{ref}}\|_\infty \le (C_g + B)\varepsilon^{\gamma - \frac{d}{2}}.$$

Choosing $\varepsilon_0$ even smaller, now also depending on $C_g$, $B$ and $\gamma$, by continuity of the stability constant, we can find a $\lambda_0 = \lambda_0(y_{\mathrm{cont}}, V, W_{\mathrm{atom}}) > 0$
such that

$$\lambda_{\mathrm{atom}}(M) \ge \lambda_0 \quad \text{and} \quad M \in V$$

for all $M \in \mathbb{R}^{d \times \mathcal{R}}$ with $|M - D_{\mathcal{R},\varepsilon}y_{\mathrm{ref}}| \le (C_g + B)\varepsilon_0^{\gamma - \frac{d}{2}}$ for any $x, t$. In
particular, we see that this is true for $M = D_{\mathcal{R},\varepsilon}y$ or $M = sD_{\mathcal{R},\varepsilon}y +
(1 - s)D_{\mathcal{R},\varepsilon}y_{\mathrm{ref}}$, $s \in [0, 1]$ as long as $t < T_\varepsilon$.

Setting

$$A_\varepsilon = \int_0^1 D^2W_{\mathrm{atom}}\big(D_{\mathcal{R},\varepsilon}y_{\mathrm{ref}} + s(D_{\mathcal{R},\varepsilon}y - D_{\mathcal{R},\varepsilon}y_{\mathrm{ref}})\big)\,ds,$$

we see that for $|x - x'| \leq 2r + 2\varepsilon R_{\max}$

$$|A_\varepsilon(x) - A_\varepsilon(x')| \leq \|D^3 W_{\text{atom}}\|_\infty \big(\|D^2 y_{\text{ref}}\|_\infty |x - x'| + 2(B + C_g)\varepsilon^{\gamma - \frac{d}{2}}\big)$$
$$\leq C(r + \varepsilon + (B + C_g)\varepsilon^{\gamma - \frac{d}{2}}).$$

If again $\varepsilon_0$ is small enough we can therefore use the atomistic Gårding inequality from Theorem 5.6 with $r = r(y_{\text{cont}}, W_{\text{atom}}, \lambda_0)$ small enough and independent of $\varepsilon$ to get

$$\mathcal{E}(t) \leq \|u\|^2_{\ell^2_\varepsilon(\text{int}_\varepsilon \Omega)} + \|\dot{u}\|^2_{\ell^2_\varepsilon(\text{int}_\varepsilon \Omega)} + \max\{2, \frac{\lambda_0}{2}\}\|u\|^2_{h^1_\varepsilon(\text{sint}_\varepsilon \Omega)}$$
$$\leq C\|u\|^2_{\ell^2_\varepsilon(\text{int}_\varepsilon \Omega)} + \max\{\frac{4}{\lambda_0}, 1\}\|\dot{u}\|^2_{\ell^2_\varepsilon(\text{int}_\varepsilon \Omega)}$$
$$+ \max\{\frac{4}{\lambda_0}, 1\}\varepsilon^d \sum_{x \in \text{sint}_\varepsilon \Omega} A_\varepsilon(x, t)[D_{\mathcal{R},\varepsilon} u]^2$$
$$\leq C\|u\|^2_{\ell^2_\varepsilon(\text{int}_\varepsilon \Omega)} + \max\{\frac{8}{\lambda_0}, 2\}$$
$$\cdot \Big(\frac{1}{2}\|\dot{u}\|^2_{\ell^2_\varepsilon(\text{int}_\varepsilon \Omega)} + \frac{1}{2}\varepsilon^d \sum_{x \in \text{sint}_\varepsilon \Omega} A_\varepsilon(x, t)[D_{\mathcal{R},\varepsilon} u]^2\Big)$$

for some $C = C(y_{\text{cont}}, W_{\text{atom}}, \lambda_0, \mathcal{R})$. If we rewrite this in terms of the initial conditions and take absolute values, we get

$$\mathcal{E}(t) \leq C\Big(\|h_{\text{atom},1} - \dot{y}_{\text{ref}}(0) - \frac{\partial}{\partial t} K_\varepsilon(y - y_{\text{ref}})(0)\|^2_{\ell^2_\varepsilon(\text{int}_\varepsilon \Omega)}$$
$$+ \|h_{\text{atom},0} - y_{\text{ref}}(0) - K_\varepsilon(y - y_{\text{ref}})(0)\|^2_{\ell^2_\varepsilon(\text{int}_\varepsilon \Omega)}$$
$$+ \|h_{\text{atom},0} - y_{\text{ref}}(0) - K_\varepsilon(y - y_{\text{ref}})(0)\|^2_{h^1_\varepsilon(\text{sint}_\varepsilon \Omega)}$$
$$+ \Big|\int_0^t (u, \dot{u})_{\ell^2_\varepsilon}\, d\tau\Big|$$
$$+ \Big|\int_0^t (\dot{u}, \ddot{u})_{\ell^2_\varepsilon} + \varepsilon^d \sum_{x \in \text{sint}_\varepsilon \Omega} A_\varepsilon(x, \tau)[D_{\mathcal{R},\varepsilon} u, D_{\mathcal{R},\varepsilon} \dot{u}]$$
$$+ \frac{1}{2}\varepsilon^d \sum_{x \in \text{sint}_\varepsilon \Omega} \dot{A}_\varepsilon(x, \tau)[D_{\mathcal{R},\varepsilon} u]^2\, d\tau\Big|\Big).$$

Using our assumptions at $t = 0$ and for the boundary conditions we can continue by

$$\mathcal{E}(t) \leq C\Big( (C_g^2 + C_h^2)\varepsilon^{2\gamma} + \int_0^t \mathcal{E}(\tau)\, d\tau$$

$$+ \Big| \int_0^t \varepsilon^d \sum_{x \in \operatorname{sint}_\varepsilon \Omega} \dot{A}_\varepsilon(x,\tau)[D_{\mathcal{R},\varepsilon} u]^2\, d\tau \Big|$$

$$+ \Big| \int_0^t \big( \dot{u}, \frac{\partial^2}{\partial t^2} K_\varepsilon (g_{\mathrm{atom}} - y_{\mathrm{ref}}) \big)_{\ell_\varepsilon^2}\, d\tau \Big|$$

$$+ \Big| \int_0^t \varepsilon^d \sum_{x \in \operatorname{sint}_\varepsilon \Omega} A_\varepsilon(x,\tau)[D_{\mathcal{R},\varepsilon} K_\varepsilon (g_{\mathrm{atom}} - y_{\mathrm{ref}}), D_{\mathcal{R},\varepsilon}\dot{u}]\, d\tau \Big|$$

$$+ \Big| \int_0^t (\dot{u}, \ddot{y} - \ddot{y}_{\mathrm{ref}})_{\ell_\varepsilon^2} + \varepsilon^d \sum_{x \in \operatorname{sint}_\varepsilon \Omega} A_\varepsilon(x,\tau)[D_{\mathcal{R},\varepsilon}(y - y_{\mathrm{ref}}), D_{\mathcal{R},\varepsilon}\dot{u}]\, d\tau \Big| \Big)$$

$$=: C\big( (C_g^2 + C_h^2)\varepsilon^{2\gamma} + \int_0^t \mathcal{E}(\tau)\, d\tau \big) + I_1 + I_2 + I_3 + I_4$$

Clearly,

$$I_2 \leq C\big( \int_0^t \mathcal{E}(\tau)\, d\tau + C_g^2 \varepsilon^{2\gamma} \big).$$

For $I_4$ we can use the estimates from the static case. Indeed, partial summation gives

$$I_4 \leq C\Big( \int_0^t \mathcal{E}(t)\, d\tau + \int_0^t \|\ddot{y} - \ddot{y}_{\mathrm{ref}}$$
$$- \operatorname{div}_{\mathcal{R},\varepsilon}(A_\varepsilon(x,\tau)D_{\mathcal{R},\varepsilon}(y - y_{\mathrm{ref}}))\|_{\ell_\varepsilon^2(\operatorname{int}_\varepsilon \Omega)}^2\, d\tau \Big)$$
$$= C\Big( \int_0^t \mathcal{E}(t)\, d\tau + \int_0^t \|\ddot{y} - \ddot{y}_{\mathrm{ref}}$$
$$- \operatorname{div}_{\mathcal{R},\varepsilon}(DW_{\mathrm{atom}}(D_{\mathcal{R},\varepsilon} y) - DW_{\mathrm{atom}}(D_{\mathcal{R},\varepsilon} y_{\mathrm{ref}}))\|_{\ell_\varepsilon^2(\operatorname{int}_\varepsilon \Omega)}^2\, d\tau \Big)$$

As we showed at the beginning of this proof, we have

$$\operatorname{co}\{ D_{\mathcal{R},\varepsilon} y_{\mathrm{ref}}(\hat{x} + \varepsilon\sigma), (\nabla y_{\mathrm{ref}}(x)\rho)_{\rho \in \mathcal{R}} \} \subset V$$

for all $x \in \Omega_\varepsilon$ and $\sigma \in \mathcal{R} \cup \{0\}$. We are therefore in a position to apply

Proposition 4.9.

$$\|\ddot{y} - \ddot{y}_{\mathrm{ref}} - \mathrm{div}_{\mathcal{R},\varepsilon}(DW_{\mathrm{atom}}(D_{\mathcal{R},\varepsilon}y) - DW_{\mathrm{atom}}(D_{\mathcal{R},\varepsilon}y_{\mathrm{ref}}))\|_{\ell^2_\varepsilon(\mathrm{int}_\varepsilon\,\Omega)}$$

$$= \|f_{\mathrm{atom}} - \ddot{y}_{\mathrm{ref}} + \mathrm{div}_{\mathcal{R},\varepsilon}\,DW_{\mathrm{atom}}(D_{\mathcal{R},\varepsilon}y_{\mathrm{ref}})\|_{\ell^2_\varepsilon(\mathrm{int}_\varepsilon\,\Omega)}$$

$$\leq \|f_{\mathrm{atom}} - \ddot{y}_{\mathrm{ref}} + \ddot{y}_{\mathrm{cont}} - \tilde{f}\|_{\ell^2_\varepsilon(\mathrm{int}_\varepsilon\,\Omega)}$$

$$+ \|-\ddot{y}_{\mathrm{cont}} + \tilde{f} + \mathrm{div}_{\mathcal{R},\varepsilon}\,DW_{\mathrm{atom}}(D_{\mathcal{R},\varepsilon}y_{\mathrm{ref}})\|_{\ell^2_\varepsilon(\mathrm{int}_\varepsilon\,\Omega)}$$

$$\leq \|f_{\mathrm{atom}} - f_{\mathrm{ref}}\|_{\ell^2_\varepsilon(\mathrm{int}_\varepsilon\,\Omega)} + \|-\ddot{y}_{\mathrm{cont}} + f + \mathrm{div}\,DW_{\mathrm{CB}}(\nabla y_{\mathrm{ref}})\|_{L^2(\Omega_\varepsilon;\mathbb{R}^d)}$$

$$+ C\varepsilon^2\Big\|\|\nabla^4 y_{\mathrm{ref}}\|_{L^\infty(B_{\varepsilon R}(x))} + \|\nabla^3 y_{\mathrm{ref}}\|_{L^\infty(B_{\varepsilon R}(x))}^{\frac{3}{2}} + \|\nabla^2 y_{\mathrm{ref}}\|_{L^\infty(B_{\varepsilon R}(x))}^3$$

$$+ \varepsilon\|\nabla^3 y_{\mathrm{ref}}\|_{L^\infty(B_{\varepsilon R}(x))}^2\Big\|_{L^2(\Omega_\varepsilon)},$$

where $C$ and $R$ just depend on $d, \mathcal{R}$ and $\|D^2 W_{\mathrm{atom}}\|_{C^2(V)}$. Now, remember that $y_{\mathrm{ref}} = \eta_\varepsilon * (E y_{\mathrm{cont}})$. Hence, we can apply Proposition 4.10 and Proposition 4.11 and get

$$\|\ddot{y} - \ddot{y}_{\mathrm{ref}} - \mathrm{div}_{\mathcal{R},\varepsilon}(DW_{\mathrm{atom}}(D_{\mathcal{R},\varepsilon}y) - DW_{\mathrm{atom}}(D_{\mathcal{R},\varepsilon}y_{\mathrm{ref}}))\|_{\ell^2_\varepsilon(\mathrm{int}_\varepsilon\,\Omega)}$$

$$\leq \|f_{\mathrm{atom}} - f_{\mathrm{ref}}\|_{\ell^2_\varepsilon(\mathrm{int}_\varepsilon\,\Omega)}$$

$$+ C\varepsilon^2\big(\|\nabla^4 E y_{\mathrm{cont}}\|_{L^2(\Omega_\varepsilon + B_{(R+1)\varepsilon}(0))} + \|\nabla^3 E y_{\mathrm{cont}}\|_{L^3(\Omega_\varepsilon + B_{(R+1)\varepsilon}(0))}^{\frac{3}{2}}$$

$$+ \|\nabla^2 E y_{\mathrm{cont}}\|_{L^6(\Omega_\varepsilon + B_{(R+1)\varepsilon}(0))}^3 + \varepsilon\|\nabla^3 E y_{\mathrm{cont}}\|_{L^4(\Omega_\varepsilon + B_{(R+1)\varepsilon}(0))}^2$$

$$+ \|\nabla^2 E y_{\mathrm{cont}}\|_{L^4(\Omega + B_\varepsilon(0))}\|\nabla^3 E y_{\mathrm{cont}}\|_{L^4(\Omega + B_\varepsilon(0))}$$

$$+ \|\nabla^4 E y_{\mathrm{cont}}\|_{L^2(\Omega + B_\varepsilon(0))}\big)$$

$$\leq \|f_{\mathrm{atom}} - f_{\mathrm{ref}}\|_{\ell^2_\varepsilon(\mathrm{int}_\varepsilon\,\Omega)} + C\varepsilon^2\|y_{\mathrm{cont}}\|_{H^4(\Omega;\mathbb{R}^d)}(1 + \|y_{\mathrm{cont}}\|_{H^4(\Omega;\mathbb{R}^d)}^2),$$

where in the last step we used standard embedding theorems with $d \in \{2,3\}$, as well as the fact that $E$ is a continuous extension operator on all Sobolev spaces. Hence, we find

$$I_4 \leq C\Big(\int_0^t \mathcal{E}(t)\,d\tau + C_f^2 \varepsilon^{2\gamma} + \varepsilon^4\Big).$$

Now let us look at the nonlinearity $I_1$. Let us first motivate the approach here. Evaluating the time derivative, we see that we can control $I_1$ in terms of some powers of the energy and $\varepsilon$. But the straightforward estimates (see further below) are not good enough in $\varepsilon$, the problematic term being of the form

$$\int_0^t \varepsilon^d \sum_{x\in\mathrm{sint}_\varepsilon\,\Omega} D^3 W_{\mathrm{atom}}\big(D_{\mathcal{R},\varepsilon}y_{\mathrm{ref}} + s(D_{\mathcal{R},\varepsilon}y - D_{\mathcal{R},\varepsilon}y_{\mathrm{ref}})\big)$$

$$[D_{\mathcal{R},\varepsilon}u, D_{\mathcal{R},\varepsilon}u, D_{\mathcal{R},\varepsilon}\dot{u}]\,d\tau.$$

Since we have the uniform estimate

$$|D_{\mathcal{R},\varepsilon}u| \leq \varepsilon^{-\frac{d}{2}} \|u\|_{h^1_\varepsilon(\mathrm{sint}_\varepsilon \Omega)} \leq B\varepsilon^{\gamma-\frac{d}{2}},$$

the estimates would improve if there where even more $D_{\mathcal{R},\varepsilon}u$-factors involved. But that is precisely what happens if we shift the time derivative back on the coefficients by partially integrating in time. So if we partially integrate in time several times, we can indeed hope for sufficiently strong estimates since they improve by $\varepsilon^{\gamma-\frac{d}{2}}$ with each step. To make these ideas precise let us extend the definition of $A_\varepsilon = A_{\varepsilon,2}$ to

$$A_{\varepsilon,k} = \int_0^1 D^k W_{\mathrm{atom}}\big(D_{\mathcal{R},\varepsilon}y_{\mathrm{ref}} + s(D_{\mathcal{R},\varepsilon}y - D_{\mathcal{R},\varepsilon}y_{\mathrm{ref}})\big)\, ds.$$

Furthermore, let us write for $k \geq 2$

$$B_k(t) = \int_0^1 \varepsilon^d \sum_{x \in \mathrm{sint}_\varepsilon \Omega} \dot{A}_{\varepsilon,k-1}[D_{\mathcal{R},\varepsilon}u]^{k-1} s^{k-3}\, ds,$$

$$C_k(t) = \int_0^1 \varepsilon^d \sum_{x \in \mathrm{sint}_\varepsilon \Omega} A_{\varepsilon,k}[D_{\mathcal{R},\varepsilon}u]^{k-1}[D_{\mathcal{R},\varepsilon}\dot{y}_{\mathrm{ref}}] s^{k-3}\, ds,$$

$$D_k(t) = \int_0^1 \varepsilon^d \sum_{x \in \mathrm{sint}_\varepsilon \Omega} A_{\varepsilon,k}[D_{\mathcal{R},\varepsilon}u]^{k-1}[D_{\mathcal{R},\varepsilon}\dot{u}] s^{k-2}\, ds,$$

$$E_k(t) = \int_0^1 \varepsilon^d \sum_{x \in \mathrm{sint}_\varepsilon \Omega} A_{\varepsilon,k}[D_{\mathcal{R},\varepsilon}u]^{k-1}[D_{\mathcal{R},\varepsilon}(K_\varepsilon(g_{\mathrm{atom}} - y_{\mathrm{ref}}))^{\cdot}] s^{k-2}\, ds,$$

$$F_k(t) = \int_0^1 \varepsilon^d \sum_{x \in \mathrm{sint}_\varepsilon \Omega} A_{\varepsilon,k}[D_{\mathcal{R},\varepsilon}u]^k s^{k-2}\, ds.$$

In this notation, we have

$$I_1 = C\left| \int_0^t B_3(\tau)\, d\tau \right|$$

and, for $3 \leq k \leq m+1$, by partial integration in time,

$$\int_0^t B_k(\tau) + (k-1)D_{k-1}(\tau)\, d\tau = F_{k-1}(t) - F_{k-1}(0),$$

as well as

$$B_k(t) = C_k(t) + D_k(t) + E_k(t)$$

by evaluating the time derivative. We claim to have relatively good estimates on the $C_k$, $E_k$, and $F_k$. At the same time we will prove estimates on the $D_k$ that get better with increasing $k$. Due to the two equations above this is sufficient. We just need to control all the $C_k$, $E_k$, and $F_k$, as well as $D_{m+1}$. Since $|D_{\mathcal{R},\varepsilon}u| \leq B\varepsilon^{\gamma - \frac{d}{2}}$ as mentioned above, we have the following estimates:

$$\left| \int_0^t C_k(\tau)\,d\tau \right| \leq C(B\varepsilon^{\gamma - \frac{d}{2}})^{k-3} \|y_{\text{cont}}\|_{C^1(0,t;H^3(\Omega))} \int_0^t \mathcal{E}(\tau)\,d\tau,$$

$$\left| \int_0^t E_k(\tau)\,d\tau \right| \leq C(B\varepsilon^{\gamma - \frac{d}{2}})^{k-2} \left( \int_0^t \mathcal{E}(\tau)\,d\tau + C_g^2 \varepsilon^{2\gamma} \right),$$

$$|F_k(0)| \leq C(B\varepsilon^{\gamma - \frac{d}{2}})^{k-2}(C_g^2 + C_h^2)\varepsilon^{2\gamma},$$

$$|F_k(t)| \leq C(B\varepsilon^{\gamma - \frac{d}{2}})^{k-2}\mathcal{E}(t).$$

Furthermore,

$$\left| \int_0^t D_{m+1}(\tau)\,d\tau \right| \leq C(B\varepsilon^{\gamma - \frac{d}{2}})^{m-1} \int_0^t \|u\|_{h_\varepsilon^1(\text{sint}_\varepsilon \Omega)} \|\dot{u}\|_{h_\varepsilon^1(\text{sint}_\varepsilon \Omega)}\,d\tau$$

$$\leq C\varepsilon^{-1}(B\varepsilon^{\gamma - \frac{d}{2}})^{m-1} \int_0^t \|u\|_{h_\varepsilon^1(\text{sint}_\varepsilon \Omega)} \|\dot{u}\|_{\ell_\varepsilon^2(\text{int}_\varepsilon \Omega)}\,d\tau$$

$$\leq C\varepsilon^{-1}(B\varepsilon^{\gamma - \frac{d}{2}})^{m-1} \int_0^t \|u\|_{h_\varepsilon^1(\text{sint}_\varepsilon \Omega)}^2 + \|\dot{u}\|_{\ell_\varepsilon^2(\text{int}_\varepsilon \Omega)}^2\,d\tau$$

$$\leq C\varepsilon^{-1}(B\varepsilon^{\gamma - \frac{d}{2}})^{m-1} \int_0^t \mathcal{E}(\tau)\,d\tau.$$

Choosing $\varepsilon_0$ small enough, such that $B\varepsilon^{\gamma - \frac{d}{2}} \leq 1$, we can drop the $B\varepsilon^{\gamma - \frac{d}{2}}$-factors wherever they are not needed and combine the estimates from $k = 3$ up to $k = m + 1$ to get

$$I_1 \leq C\Big( \int_0^t \mathcal{E}(\tau)\,d\tau + (C_g^2 + C_h^2)\varepsilon^{2\gamma} + B\varepsilon^{\gamma - \frac{d}{2}}\mathcal{E}(t)$$

$$+ B^{m-1}\varepsilon^{(m-1)(\gamma - \frac{d}{2})-1} \int_0^t \mathcal{E}(\tau)\,d\tau \Big)$$

for some $C = C(y_{\text{cont}}, W_{\text{atom}}, V, \mathcal{R}, \Omega, m)$. Choosing $\varepsilon_0$ even smaller, we can ensure that $CB\varepsilon^{\gamma - \frac{d}{2}} \leq \frac{1}{3}$ and $B^{m-1}\varepsilon^{(m-1)(\gamma - \frac{d}{2})-1} \leq 1$, since $\gamma > \frac{d}{2} + \frac{1}{m-1}$ by assumption. Therefore,

$$I_1 \leq \frac{1}{3}\mathcal{E}(t) + C\Big( \int_0^t \mathcal{E}(\tau)\,d\tau + (C_g^2 + C_h^2)\varepsilon^{2\gamma} \Big).$$

This scheme can also be adapted to deal with the additional error term $I_3$ coming from the boundary conditions. We now set

$$B_k(t) = \int_0^1 \varepsilon^d \sum_{x \in \text{sint}_\varepsilon \Omega} \dot{A}_{\varepsilon,k-1}[D_{\mathcal{R},\varepsilon} K_\varepsilon(g_{\text{atom}} - y_{\text{ref}})][D_{\mathcal{R},\varepsilon} u]^{k-2} s^{k-3} \, ds,$$

$$C_k(t) = \int_0^1 \varepsilon^d \sum_{x \in \text{sint}_\varepsilon \Omega} A_{\varepsilon,k}[D_{\mathcal{R},\varepsilon} K_\varepsilon(g_{\text{atom}} - y_{\text{ref}})][D_{\mathcal{R},\varepsilon} u]^{k-2}$$
$$[D_{\mathcal{R},\varepsilon} \ddot{y}_{\text{ref}}] s^{k-3} \, ds,$$

$$D_k(t) = \int_0^1 \varepsilon^d \sum_{x \in \text{sint}_\varepsilon \Omega} A_{\varepsilon,k}[D_{\mathcal{R},\varepsilon} K_\varepsilon(g_{\text{atom}} - y_{\text{ref}})][D_{\mathcal{R},\varepsilon} u]^{k-2}$$
$$[D_{\mathcal{R},\varepsilon} \dot{u}] s^{k-2} \, ds,$$

$$E_k(t) = \int_0^1 \varepsilon^d \sum_{x \in \text{sint}_\varepsilon \Omega} A_{\varepsilon,k}[D_{\mathcal{R},\varepsilon} K_\varepsilon(g_{\text{atom}} - y_{\text{ref}})][D_{\mathcal{R},\varepsilon} u]^{k-2}$$
$$[D_{\mathcal{R},\varepsilon}(K_\varepsilon(g_{\text{atom}} - y_{\text{ref}}))\dot{}] s^{k-2} \, ds,$$

$$F_k(t) = \int_0^1 \varepsilon^d \sum_{x \in \text{sint}_\varepsilon \Omega} A_{\varepsilon,k}[D_{\mathcal{R},\varepsilon} K_\varepsilon(g_{\text{atom}} - y_{\text{ref}})][D_{\mathcal{R},\varepsilon} u]^{k-1} s^{k-2} \, ds$$

$$G_k(t) = \int_0^1 \varepsilon^d \sum_{x \in \text{sint}_\varepsilon \Omega} A_{\varepsilon,k}[D_{\mathcal{R},\varepsilon}(K_\varepsilon(g_{\text{atom}} - y_{\text{ref}}))\dot{}][D_{\mathcal{R},\varepsilon} u]^{k-1} s^{k-2} \, ds.$$

In analogy to before, we have

$$I_3 = C \left| \int_0^t D_2(\tau) \, d\tau \right|$$

and, for $3 \leq k \leq m + 1$,

$$\int_0^t B_k(\tau) + (k-2)D_{k-1}(\tau) + G_{k-1}(\tau) \, d\tau = F_{k-1}(t) - F_{k-1}(0),$$

as well as

$$B_k(t) = C_k(t) + D_k(t) + E_k(t).$$

Again, we have the estimates

$$\left| \int_0^t C_k(\tau)\, d\tau \right| \le C(B\varepsilon^{\gamma-\frac{d}{2}})^{k-3} \|y_{\mathrm{cont}}\|_{C^1(0,t;H^3(\Omega))}$$

$$\cdot \left( \int_0^t \mathcal{E}(\tau)\, d\tau + C_g^2 \varepsilon^{2\gamma} \right),$$

$$\left| \int_0^t E_k(\tau)\, d\tau \right| \le C(B\varepsilon^{\gamma-\frac{d}{2}})^{k-2} C_g^2 \varepsilon^{2\gamma},$$

$$\left| \int_0^t G_k(\tau)\, d\tau \right| \le C(B\varepsilon^{\gamma-\frac{d}{2}})^{k-2} \left( \int_0^t \mathcal{E}(\tau)\, d\tau + C_g^2 \varepsilon^{2\gamma} \right),$$

$$|F_k(0)| \le C(B\varepsilon^{\gamma-\frac{d}{2}})^{k-2}(C_g^2 + C_h^2)\varepsilon^{2\gamma},$$

$$|F_k(t)| \le C(B\varepsilon^{\gamma-\frac{d}{2}})^{k-2}(\mathcal{E}(t) + C_g^2 \varepsilon^{2\gamma}).$$

Furthermore,

$$\left| \int_0^t D_{m+1}(\tau)\, d\tau \right| \le C\varepsilon^{-1}(B\varepsilon^{\gamma-\frac{d}{2}})^{m-1} \left( \int_0^t \mathcal{E}(\tau)\, d\tau + C_g^2 \varepsilon^{2\gamma} \right).$$

As before this implies

$$I_3 \le \frac{1}{3}\mathcal{E}(t) + C\left( \int_0^t \mathcal{E}(\tau)\, d\tau + (C_g^2 + C_h^2)\varepsilon^{2\gamma} \right)$$

for $\varepsilon$ small enough.

Overall we have thus proven

$$\mathcal{E}(t) = 3\left(\mathcal{E}(t) - \frac{2}{3}\mathcal{E}(t)\right)$$

$$\le C\left( (C_f^2 + C_g^2 + C_h^2 + \varepsilon^{4-2\gamma})\varepsilon^{2\gamma} + \int_0^t \mathcal{E}(\tau)\, d\tau \right)$$

for some $C = C(y_{\mathrm{cont}}, W_{\mathrm{atom}}, V, \mathcal{R}, \Omega, m, \gamma)$, all $\varepsilon$ with $0 < \varepsilon \le \varepsilon_0(y_{\mathrm{cont}}, W_{\mathrm{atom}}, V, \mathcal{R}, \Omega, m, \gamma, B)$ and $t \in [0, T_\varepsilon)$.

Grönwall's inequality then yields

$$\mathcal{E}(t) \le C(C_f^2 + C_g^2 + C_h^2 + \varepsilon^{4-2\gamma})\varepsilon^{2\gamma} e^{Ct}$$

$$\le \frac{B^2}{2}\varepsilon^{2\gamma},$$

where we have finally chosen $B := \left( 2C(C_f^2 + C_g^2 + C_h^2 + 1)e^{CT_0} \right)^{\frac{1}{2}}$. In particular, with $C \ge 1$ we satisfy the condition $B > \sqrt{2C_g^2 + 2C_h^2}$, that we required at the beginning.

Most notably, we find $\mathcal{E}(T_\varepsilon) \leq \frac{B^2}{2}\varepsilon^{2\gamma}$ and therefore $T_\varepsilon = T_0$ for $\varepsilon \leq \varepsilon_0$. The claimed convergence estimate immediately follows from the energy estimate above and the estimate we assumed for the boundary conditions. Uniqueness follows directly from the local uniqueness in Theorem 5.5.                                                                                      □

# Chapter 6

# A Complimentary Convergence Result

We want to complement the existence and convergence results in the static case from Chapter 4 with a pure convergence result in the same setting that does not say anything about existence but has significantly weaker assumptions. The proof is more elegant and not as involved, which highlights the fact that the existence of the atomistic solutions with the desired properties (and their convergence rate) are really the difficult and most interesting parts of Theorem 4.5.

Let us be more precise on what we will do here. We will start with a sequence of stable atomistic deformations that solve the atomistic problem and from there we will prove that this sequence converges strongly in $H^1$ (no subsequences are needed), that the limit is a solution of the Cauchy-Born problem, and, additionally, that the energies converge. All of this will be done with very minimal assumptions on body forces and boundary values.

Even though the proof uses some variational ideas in the spirit of $\Gamma$-convergence, it should be pointed out that there is a significant mismatch in topologies between $H^1$ and $W^{1,\infty}$ that prevents any (straightforward) extension to even a local $\Gamma$-convergence result. Besides the variational ideas, the main motivation for the following result comes from the theory of (uniformly) monotone operators which might become more clear in the proof. Despite these motivations, it should be remarked that the operators representing the equations are, typically, not monotone and that the elastic energy density is typically not quasiconvex.

**Theorem 6.1.** *Let $d \geq 2$ and let $\Omega \subset \mathbb{R}^d$ be open, bounded, and with Lipschitz boundary. Let $R > 0$, $A_0 \in \mathbb{R}^{d \times d}$, take $\mathcal{R}$ as before, let*

149

$W_{\text{atom}} \in C^2(B_R((A_0\rho)_{\rho\in\mathcal{R}}))$, and assume $\lambda_{\text{atom}}(A_0) > 0$. Then there are $r_1, r_2 > 0$ such that the following holds:

Given maps $g \in W^{1,\infty}_{\text{loc}}(\mathbb{R}^d, \mathbb{R}^d)$ and $f \in L^1(\Omega; \mathbb{R}^d)$, as well as sequences $\varepsilon_n \to 0$, $y_n : \Omega \cap \varepsilon_n \mathbb{Z}^d \to \mathbb{R}^d$, $g_n : \partial_\varepsilon \Omega \to \mathbb{R}^d$, and $f_n : \text{int}_{\varepsilon_n} \Omega \to \mathbb{R}^d$, such that

$$|g_n(x) - g(x)| \le r_1\varepsilon_n \quad \text{for all } x \in \partial_{\varepsilon_n}\Omega, n \in \mathbb{N},$$
$$|\nabla g(x) - A_0| \le r_2 \quad \text{in a neighborhood of } \partial\Omega,$$
$$|D_{\mathcal{R},\varepsilon_n}y_n(x) - (A_0\rho)_{\rho\in\mathcal{R}}| \le r_2 \quad \text{for all } x \in \text{sint}_{\varepsilon_n}\Omega, n \in \mathbb{N},$$
$$f_n \rightharpoonup f \quad \text{in } L^1(\Omega; \mathbb{R}^d),$$

and such that $y_n$ solves the atomistic equations with body forces $f_n$ subject to the boundary values $g_n$. Then there is a $y \in W^{1,\infty}(\Omega; \mathbb{R}^d)$ such that $y_n \to y$ uniformly, $D_{\mathcal{R},\varepsilon_n}y_n \to (\nabla y\rho)_{\rho\in\mathcal{R}}$ strongly in $L^2(\Omega; \mathbb{R}^{d\times\mathcal{R}})$ and weak-$*$ in $L^\infty(\Omega; \mathbb{R}^{d\times\mathcal{R}})$, and $E_{\varepsilon_n}(y_n; f_n; g_n) \to E(y; f; g)$. Furthermore, $y$ is a weak solution of the continuous equations with boundary values $g$ and body forces $f$ as well as a local minimizer of $E(\cdot; f; g)$ in $W^{1,\infty}(\Omega; \mathbb{R}^d)$ (uniform in the $H^1_0(\Omega; \mathbb{R}^d)$-norm).

*Remark* 6.2. To make the convergence precise, wherever necessary atomistic functions are identified with the functions that are constant on every cube $Q_\varepsilon$ (or more precisely, on the corresponding half open cubes). Furthermore, for the sake of definiteness we set $f_n = 0$ outside of $\bigcup_{z\in\text{int}_\varepsilon \Omega} Q_\varepsilon(z)$, $D_{\mathcal{R},\varepsilon_n}y_n = 0$ outside of $\bigcup_{z\in\text{sint}_\varepsilon \Omega} Q_\varepsilon(z)$, and $y_n = g$ outside of $\bigcup_{z\in\Omega\cap\varepsilon\mathbb{Z}^d} Q_\varepsilon(z)$.

*Proof.* The proof is build up as follows: First, we argue by compactness to get weak convergence on a subsequence to a limit $y$. Then, we construct a "recovery sequence" for $y$ with better convergence properties but the same boundary conditions as the original sequence. As a third step we compare the energies along these sequences in two different ways which will allow us to prove strong convergence and convergence of the energies for the subsequence we are looking at. With the help of strong convergence we can pass to the limit in the nonlinear Euler-Lagrange equations. As a last step we show that the limit $y$ is a local minimizer and as such uniquely determined, so that actually the entire sequence converges.

Step 1: compactness.

Since $g_n$ and $D_{\mathcal{R},\varepsilon_n}y_n$ are bounded independently of $n$ and $x$ the $y_n$ are bounded uniformly in $\Omega$, as we can write for some fixed $\rho \in \mathcal{R}$

$$y_n(x) = g_n(x + (L(x,n) + 1)\rho) + \sum_{j=0}^{L(x,n)} y_n(x + j\rho) - y_n(x + (j+1)\rho),$$

with $|L(x,n)\varepsilon_n| \leq C$. Since $\mathrm{span}_{\mathbb{Z}} \mathcal{R} = \mathbb{Z}^d$, we can fix specific representations for the vectors $\pm e_i$, where $(e_i)_i$ is the standard basis of $\mathbb{R}^d$. In particular, we find

$$|y_n(x \pm \varepsilon_n e_i) - y_n(x)| \leq C\varepsilon_n,$$

whenever $\mathrm{dist}(x, \partial\Omega) \geq R\varepsilon_n$ for $R$ large enough. Additionally, we know that $\partial\Omega$ is Lipschitz, $\nabla g$ is bounded, and $|y_n - g| \leq C\varepsilon_n$ on $\partial_{\varepsilon_n}\Omega$. Combining all of this one finds

$$|y_n(x) - y_n(x')| \leq C(\varepsilon_n + |x - x'|)$$

for all $x, x' \in \Omega$ and $n$ sufficiently large. Clearly, this is enough for an Arzelà-Ascoli-type argument, even though the maps are not continuous. So we have some subsequence (not relabeled) and a $y \in W^{1,\infty}(\Omega, \mathbb{R}^d)$, such that $y_n \to y$ uniformly. On the boundary, we directly see $y = g$. By weak-$*$-compactness in $L^\infty(\Omega; \mathbb{R}^{d\times\mathcal{R}})$, we also have some $M \in L^\infty(\Omega; \mathbb{R}^{d\times\mathcal{R}})$ with $D_{\mathcal{R},\varepsilon_n} y_n \overset{*}{\rightharpoonup} M$. To identify this limit, we calculate with $\varphi \in C_c^\infty(\Omega)$

$$\int_\Omega (M(x))_{j\rho}\varphi(x)\, dx = \lim_{n\to\infty} \int_\Omega \frac{y_n(x + \varepsilon_n\rho)_j - y_n(x)_j}{\varepsilon_n}\varphi(x)\, dx$$

$$= -\lim_{n\to\infty} \int_\Omega y_n(x)_j \frac{\varphi(x) - \varphi(x - \varepsilon_n\rho)}{\varepsilon_n}\, dx$$

$$= -\int_\Omega y(x)_j \nabla\varphi(x)\rho\, dx.$$

Which shows that $M(x) = (\nabla y(x)\rho)_{\rho\in\mathcal{R}}$ almost everywhere. As a corollary we also get

$$|(\nabla y\rho)_{\rho\in\mathcal{R}} - (A_0\rho)_{\rho\in\mathcal{R}}| \leq r_2 \quad \text{for almost all } x \in \Omega.$$

Step 2: recovery sequence.
We define a recovery sequence just by setting

$$\tilde{y}_n(x) = \begin{cases} \fint_{Q_{\varepsilon_n}(x)} y(\tilde{x})\, d\tilde{x} & \text{if } x \in \mathrm{int}_{\varepsilon_n}\Omega, \\ g_n(x) & x \in \partial_{\varepsilon_n}\Omega. \end{cases}$$

Let us check that the recovery sequence satisfies the constraint for the gradient. For $x \in \text{int}_\varepsilon \Omega$ with $x + \varepsilon_n \mathcal{R} \subset \text{int}_\varepsilon \Omega$ we have

$$\left| D_{\mathcal{R},\varepsilon_n} \fint_{Q_{\varepsilon_n}(x)} y(\tilde{x}) \, d\tilde{x} - (A_0\rho)_{\rho \in \mathcal{R}} \right|^2$$

$$= \sum_\rho \left| \fint_{Q_{\varepsilon_n}(x)} \frac{y(\tilde{x} + \varepsilon_n \rho) - y(\tilde{x})}{\varepsilon_n} - A_0\rho \, d\tilde{x} \right|^2$$

$$\leq \int_0^1 \fint_{Q_{\varepsilon_n}(x)} \sum_\rho |(\nabla y(\tilde{x} + t\varepsilon_n \rho) - A_0)\rho|^2 \, d\tilde{x} \, dt$$

$$\leq r_2^2.$$

But close to the boundary $\tilde{y}_n$ is defined differently. Here the estimates cannot be that sharp, but we claim that we still have

$$|D_{\mathcal{R},\varepsilon_n} \tilde{y}_n(x) - (A_0\rho)_{\rho \in \mathcal{R}}| \leq C(r_1 + r_2),$$

with a $C = C(\mathcal{R}, d)$. To prove this let us first look at $x \in \text{int}_{\varepsilon_n} \Omega$ with $x + \varepsilon_n \rho \in \partial_{\varepsilon_n} \Omega$. Let us fix a $z \in \partial\Omega$ with $|x - z| \leq C\varepsilon_n$ and estimate

$$\left| \frac{g_n(x + \varepsilon_n \rho) - \fint_{Q_{\varepsilon_n}(x)} y(\tilde{x}) \, d\tilde{x}}{\varepsilon_n} - A_0\rho \right|$$

$$\leq r_1 + \left| \frac{g(x + \varepsilon_n \rho) - g(z) - \fint_{Q_{\varepsilon_n}(x)} y(\tilde{x}) - y(z) \, d\tilde{x}}{\varepsilon_n} - A_0\rho \right|$$

$$\leq r_1 + \frac{1}{\varepsilon_n} \int_0^1 \fint_{Q_{\varepsilon_n}(x)} \big| (\nabla g(z + t(x + \varepsilon_n \rho - z)) - A_0)[x - z + \varepsilon_n \rho]$$

$$- (\nabla y(z + t(\tilde{x} - z)) - A_0)[\tilde{x} - z] \big| \, dt \, d\tilde{x}$$

$$\leq r_1 + Cr_2.$$

The same arguments hold if $x$ and $x + \varepsilon_n \rho$ are interchanged. If both points are in $\text{int}_{\varepsilon_n} \Omega$ we can argue as before. Finally, if $x, x + \varepsilon_n \rho \in \partial_{\varepsilon_n} \Omega$ we have

$$\left| \frac{g_n(x + \varepsilon_n \rho) - g_n(x)}{\varepsilon_n} - A_0\rho \right|$$

$$\leq 2r_1 + \left| \frac{g(x + \varepsilon_n \rho) - g(x)}{\varepsilon_n} - A_0\rho \right|$$

$$\leq 2r_1 + \int_0^1 \big| (\nabla g(x + t\varepsilon_n \rho) - A_0)\rho \big| \, dt$$

$$\leq 2r_1 + Cr_2.$$

Therefore,

$$|D_{\mathcal{R},\varepsilon_n}\tilde{y}_n(x) - (A_0\rho)_{\rho\in\mathcal{R}}| \leq C(r_1 + r_2)$$

for all $x \in \mathrm{sint}_{\varepsilon_n}\Omega$ and all $n$.

Before we choose $r_1$ and $r_2$, remember, that we showed in the proof of Theorem 4.5 that the positivity of $\lambda_{\mathrm{atom}}(A_0)$ combined with the continuity of $D^2 W_{\mathrm{atom}}$ implies the existence of a positive $\tilde{R} < R$ such that

$$\varepsilon^d \sum_{x\in\mathrm{sint}_\varepsilon\Omega} D^2 W_{\mathrm{atom}}(D_{\mathcal{R},\varepsilon}w(x))[D_{\mathcal{R},\varepsilon}u(x)]^2$$
$$\geq \frac{\lambda_{\mathrm{atom}}(A_0)}{2}\varepsilon^d \sum_{x\in\mathrm{sint}_\varepsilon\Omega} |D_{\mathcal{R},\varepsilon}u(x)|^2$$

for all $u \in \mathcal{A}_\varepsilon(\Omega,0)$ and all $w\colon \Omega \cap \varepsilon\mathbb{Z}^d \to \mathbb{R}^d$ with

$$|D_{\mathcal{R},\varepsilon}w(x) - (A_0\rho)_{\rho\in\mathcal{R}}| \leq \tilde{R}$$

for all $x \in \mathrm{sint}_\varepsilon\Omega$. Furthermore, we can also ensure that we have

$$\int_\Omega D^2 W_{\mathrm{CB}}(\nabla w(x))[\nabla u(x)]^2\,dx \geq \frac{\lambda_{\mathrm{atom}}(A_0)}{2}\int_\Omega |\nabla u(x)|^2\,dx$$

for all $u \in H_0^1(\Omega;\mathbb{R}^d)$ and all $w \in W^{1,\infty}(\Omega;\mathbb{R}^d)$ with

$$|\nabla w(x) - A_0| \leq \tilde{R}$$

for almost all $x \in \Omega$. We now choose $r_1$ and $r_2$ small enough such that $C(r_1 + r_2) \leq \tilde{R} < R$.

Clearly we have $\tilde{y}_n \to y$ uniformly. We claim that we also have strong convergence of the difference quotients in, e.g., the $L^2$-norm. Because of the uniform boundedness, it suffices to prove pointwise convergence almost everywhere. Indeed, for every $x$ with $\mathrm{dist}(x,\Omega^c) > C\varepsilon_n$, with $C$ large enough, we find

$$|D_{\mathcal{R},\varepsilon_n}\tilde{y}_n(x) - (\nabla y(x)\rho)_{\rho\in\mathcal{R}}|^2$$
$$= \sum_\rho \left|\fint_{Q_{\varepsilon_n}(x)} \frac{y(z+\varepsilon_n\rho) - y(z)}{\varepsilon_n}\,dz - \nabla y(x)\rho\right|^2$$
$$\leq 2\sum_\rho \left|\fint_{Q_{\varepsilon_n}(x)} \frac{y(z+\varepsilon_n\rho) - y(z)}{\varepsilon_n} - \nabla y(z)\rho\,dz\right|^2$$
$$+ 2\sum_\rho \left|\fint_{Q_{\varepsilon_n}(x)} \nabla y(z)\rho\,dz - \nabla y(x)\rho\right|^2.$$

It is well known that the second term goes to zero for every Lebesgue point of $\nabla y \in L^\infty(\Omega; \mathbb{R}^d)$ (even though the cubes are not centered). The first term is slightly more difficult. Note that the integrand $\varphi_n(z) = \frac{y(z+\varepsilon_n\rho)-y(z)}{\varepsilon_n} - \nabla y(z)\rho$ is uniformly bounded and converges pointwise almost everywhere to 0 by Rademacher's theorem. These properties of $\varphi_n$ are sufficient to get the desired convergence. E.g., we can argue as follows. Since $\Omega$ is bounded, the $\varphi_n$ also converge almost uniformly. For every $\delta > 0$ we can thus find a measurable set $A_\delta$ with $|A_\delta| \leq \delta$ such that $\varphi_n \to 0$ uniformly on $A_\delta^c$. Furthermore, for almost every $x \in A_\delta^c$ we have

$$\frac{|A_\delta \cap Q_\varepsilon(x)|}{|Q_\varepsilon(x)|} \to 0.$$

For any such $x$ we find

$$\left| \fint_{Q_\varepsilon(x)} \varphi_n(z)\,dz \right| \leq \frac{|A_\delta \cap Q_\varepsilon(x)|}{|Q_\varepsilon(x)|} \|\varphi_n\|_\infty + \|\varphi_n|_{A_\delta^c}\|_\infty \to 0.$$

We thus have

$$\fint_{Q_\varepsilon(x)} \varphi_n(z)\,dz \to 0$$

for all $x$ except on a set of measure at most $\delta$. But $\delta > 0$ was arbitrary which concludes the proof of the claim.

Step 3: comparison of the energies.

On the one hand we have

$$
\begin{aligned}
E_{\varepsilon_n}(\tilde{y}_n; f_n; g_n) &- E_{\varepsilon_n}(y_n; f_n; g_n) \\
&= \int_0^1 (1-t)\varepsilon_n^d \sum_{x \in \mathrm{sint}_{\varepsilon_n}\Omega} D^2 W_{\mathrm{atom}}(D_{\mathcal{R},\varepsilon_n}(y_n + t(\tilde{y}_n - y_n))) \\
&\qquad [D_{\mathcal{R},\varepsilon_n}(\tilde{y}_n - y_n)]^2\,dt \\
&\geq \frac{\lambda_{\mathrm{atom}}(A_0)}{4} \varepsilon_n^d \sum_{x \in \mathrm{sint}_{\varepsilon_n}\Omega} |D_{\mathcal{R},\varepsilon_n}(\tilde{y}_n - y_n)|^2.
\end{aligned}
$$

On the other hand,

$$
\begin{aligned}
E_{\varepsilon_n}(y_n; f_n; g_n) &- E_{\varepsilon_n}(\tilde{y}_n; f_n; g_n) + \varepsilon_n^d \sum_{x \in \mathrm{int}_{\varepsilon_n}\Omega} f_n(y_n - \tilde{y}_n) \\
&- \varepsilon_n^d \sum_{x \in \mathrm{sint}_{\varepsilon_n}\Omega} DW_{\mathrm{atom}}(D_{\mathcal{R},\varepsilon_n}\tilde{y}_n)[D_{\mathcal{R},\varepsilon_n}(y_n - \tilde{y}_n)]
\end{aligned}
$$

$$= \int_0^1 (1-t)\varepsilon_n^d \sum_{x \in \mathrm{sint}_{\varepsilon_n} \Omega} D^2 W_{\mathrm{atom}}(D_{\mathcal{R},\varepsilon_n}(\tilde{y}_n + t(y_n - \tilde{y}_n)))$$
$$[D_{\mathcal{R},\varepsilon_n}(y_n - \tilde{y}_n)]^2$$
$$\geq \frac{\lambda_{\mathrm{atom}}(A_0)}{4} \varepsilon_n^d \sum_{x \in \mathrm{sint}_{\varepsilon_n} \Omega} |D_{\mathcal{R},\varepsilon_n}(\tilde{y}_n - y_n)|^2.$$

Combining weak convergence in $L^1(\Omega; \mathbb{R}^d)$ with strong convergence in $L^\infty(\Omega; \mathbb{R}^d)$ we surely find

$$\varepsilon_n^d \sum_{x \in \mathrm{int}_{\varepsilon_n} \Omega} f_n(y_n - \tilde{y}_n) = \int_\Omega f_n(y_n - \tilde{y}_n) \, dx \to 0.$$

Similarly, combining strong and weak convergence in $L^2(\Omega; \mathbb{R}^{d \times \mathcal{R}})$ we find

$$\varepsilon_n^d \sum_{x \in \mathrm{sint}_{\varepsilon_n} \Omega} DW_{\mathrm{atom}}(D_{\mathcal{R},\varepsilon_n}\tilde{y}_n)[D_{\mathcal{R},\varepsilon_n}(y_n - \tilde{y}_n)] \to 0.$$

Additionally, we also have $E_{\varepsilon_n}(\tilde{y}_n; f_n; g_n) \to E(y; f; g)$. Combining these limits with the two inequalities above we get $E_{\varepsilon_n}(y_n; f_n; g_n) \to E(y; f; g)$ and $D_{\mathcal{R},\varepsilon_n}(\tilde{y}_n - y_n) \to 0$ strongly in $L^2(\Omega; \mathbb{R}^{d \times \mathcal{R}})$. Using the already established strong convergence of $D_{\mathcal{R},\varepsilon_n}\tilde{y}_n$, we see that also $D_{\mathcal{R},\varepsilon_n}y_n \to (\nabla y(x)\rho)_{\rho \in \mathcal{R}}$ strongly in $L^2(\Omega; \mathbb{R}^{d \times \mathcal{R}})$. Taking another subsequence if necessary, we even find $D_{\mathcal{R},\varepsilon_n}y_n(x) \to (\nabla y(x)\rho)_{\rho \in \mathcal{R}}$ pointwise almost everywhere.

Step 4: conclusion.

Having proven the stronger convergence, we can now take the limit in the (weak) Euler-Lagrange equations. Let $\varphi \in C_c^\infty(\Omega; \mathbb{R}^d)$, then we have for all $n$ sufficiently large

$$0 = \varepsilon_n^d \sum_{x \in \mathrm{int}_{\varepsilon_n} \Omega} (-\mathrm{div}_{\mathcal{R},\varepsilon_n} DW_{\mathrm{atom}}(D_{\mathcal{R},\varepsilon_n}y_n(x)) - f_n(x))\varphi(x)$$
$$= \varepsilon_n^d \sum_{x \in \mathrm{int}_{\varepsilon_n} \Omega} DW_{\mathrm{atom}}(D_{\mathcal{R},\varepsilon_n}y_n(x))[D_{\mathcal{R},\varepsilon_n}\varphi(x)] - f_n(x)\varphi(x)$$
$$\to \int_\Omega DW_{\mathrm{atom}}((\nabla y(x)\rho)_{\rho \in \mathcal{R}})[(\nabla \varphi(x)\rho)_{\rho \in \mathcal{R}}] - f(x)\varphi(x) \, dx$$
$$= \int_\Omega DW_{\mathrm{CB}}(\nabla y(x))[\nabla \varphi(x)] - f(x)\varphi(x) \, dx.$$

$y$ also is a local minimizer since

$$E(y + u; f; g) - E(y; f; g)$$
$$= \int_\Omega \int_0^1 (1 - t) D^2 W_{\mathrm{CB}}(\nabla y + t\nabla u)[\nabla u]^2 \, dt \, dx$$
$$\geq \frac{\lambda_{\mathrm{atom}}(A_0)}{4} \int_\Omega |\nabla u|^2 \, dx,$$

whenever $u \in H_0^1(\Omega; \mathbb{R}^d) \cap W^{1,\infty}(\Omega; \mathbb{R}^d)$ with $\|\nabla u\|_{L^\infty}$ small enough. If $r_2$ is small enough, this inequality ensures that $y$ is uniquely determined as the global minimizer subject to the boundary condition $g$ and the constraint

$$|(\nabla y \rho)_{\rho \in \mathcal{R}} - (A_0 \rho)_{\rho \in \mathcal{R}}| \leq r_2 \quad \text{for almost all } x \in \Omega.$$

But if $y$ has a unique characterization independent of the subsequence considered, then a standard argument about subsubsequences shows the desired convergence for the entire sequence $(y_n)_{n \in \mathbb{N}}$ and not just a subsequence. $\qquad\qquad\square$

# Appendix A

# Additional Regularity of Weak Solutions to Second Order Hyperbolic Equations.

**Lemma A.1.** *Let $V$ be a real reflexive Banach space. Let $T > 0$ and $A, A' \colon [0,T] \to L(V; V')$ be bounded such that for every $v_1, v_2 \in V$ the map $t \mapsto \langle A(t)v_1, v_2 \rangle_{V',V}$ is absolutely continuous with derivative $\langle A'(t)v_1, v_2 \rangle$. Furthermore, assume that $A(t)$ is symmetric. Let $u \in W^{1,1}(0,T;V)$. Then*

$$g \colon t \mapsto \langle A(t)u(t), u(t) \rangle_{V',V}$$

*is absolutely continuous with derivative*

$$g'(t) = \langle A'(t)u(t), u(t) \rangle_{V',V} + 2\langle A(t)u(t), u'(t) \rangle_{V',V}.$$

*Proof.* Let $0 \le t_0 < t_1 \le T$. For $k \in \mathbb{N}$ we write $\varepsilon = \frac{t_1 - t_0}{k}$ and for any map $f$ defined on $[t_0, t_1]$ we set

$$\tilde{f}_k(t) = f\big(t_0 + (l + \tfrac{1}{2})\varepsilon\big) \qquad \text{if } t \in t_0 + [l\varepsilon, (l+1)\varepsilon)$$

$$f_k^+(t) = f\big(t_0 + (l + \tfrac{1}{2})\varepsilon\big) \qquad \text{if } t \in t_0 + [(l - \tfrac{1}{2}\varepsilon, (l + \tfrac{1}{2})\varepsilon)$$

$$f_k^\circ(t) = f\big(t_0 + l\varepsilon\big) \qquad \text{if } t \in t_0 + [(l - \tfrac{1}{2}\varepsilon, (l + \tfrac{1}{2})\varepsilon)$$

$$f_k^-(t) = f\big(t_0 + (l - \tfrac{1}{2})\varepsilon\big) \qquad \text{if } t \in t_0 + [(l - \tfrac{1}{2}\varepsilon, (l + \tfrac{1}{2})\varepsilon)$$

whenever $t \in t_0 + [l\varepsilon, (l+1)\varepsilon)$. With this notation one easily calculates

$$\int_{t_0}^{t_1} \langle A'(t)\tilde{u}_k(t), \tilde{u}_k(t) \rangle_{V',V} \, dt = -2 \int_{t_0+\frac{\varepsilon}{2}}^{t_1-\frac{\varepsilon}{2}} \langle A_k^\circ(t)u(t), u'(t) \rangle_{V',V} \, dt$$

$$\langle A(t_1)u_k^-(t_1), u_k^-(t_1) \rangle_{V',V} - \langle A(t_0)u_k^+(t_0), u_k^+(t_0) \rangle_{V',V}$$

$$- 2 \int_{t_0+\frac{\varepsilon}{2}}^{t_1-\frac{\varepsilon}{2}} \langle A_k^\circ(t) \big( \frac{u_k^+(t) + u_k^-(t)}{2} - u(t) \big), u'(t) \rangle_{V',V} \, dt.$$

Since $\|u'\| \in L^1(0,T)$, standard theory of absolutely continuous functions gives

$$\Big\| \frac{u_k^+ + u_k^-}{2} - v \Big\|_{L^\infty(t_0+\frac{\varepsilon}{2}, t_1-\frac{\varepsilon}{2}; V)} \to 0$$

as $k \to \infty$. Since also $A_k^\circ$ is uniformly bounded, the last integral goes to 0 as $k \to \infty$. The middle terms converge to $\langle A(t_1)u(t_1), u(t_1) \rangle_{V',V} - \langle A(t_0)u(t_0), u(t_0) \rangle_{V',V}$ due to the continuity of $u$. The integrand in the first term of the right hand side (extended by 0 if necessary) converges pointwise, since $s \mapsto \langle A(s)u(t), u'(t) \rangle_{V',V}$ is continuous. Therefore the limit function is measurable and the integral converges by Lebesgue's theorem to

$$\int_{t_0}^{t_1} \langle A(t)u(t), u'(t) \rangle_{V',V} \, dt.$$

The left hand side, also converges by Lebesgue's theorem. In particular, the map $t \mapsto \langle A(t)u(t), u'(t) \rangle_{V',V}$ is measurable and

$$\int_{t_0}^{t_1} \langle A(t)u(t), u'(t) \rangle_{V',V} + 2\langle A(t)u(t), u'(t) \rangle_{V',V} \, dt$$
$$= \langle A(t_1)u(t_1), u(t_1) \rangle_{V',V} - \langle A(t_0)u(t_0), u(t_0) \rangle_{V',V}$$

$\square$

**Theorem A.2.** *Let $V \hookrightarrow H \hookrightarrow V'$ be a real Gelfand triple with a reflexive Banach space $V$. Let $T > 0$ and $A, A' : [0,T] \to L(V; V')$ be bounded such that for every $v_1, v_2 \in V$ the map $t \mapsto \langle A(t)v_1, v_2 \rangle_{V',V}$ is absolutely continuous with derivative $\langle A'(t)v_1, v_2 \rangle$. Furthermore, assume that $A(t)$ is symmetric and satisfies the uniform Gårding inequality*

$$\langle A(t)v, v \rangle_{V',V} \geq c_1 \|v\|_V^2 - c_2 \|v\|_H^2$$

*for some constants $c_1 > 0$, $c_2 \in \mathbb{R}$ and all $t, v$. Let $u \in L^\infty(0,T; V)$ such that the weak derivatives $u' \in L^\infty(0,T; H)$ and $u'' \in L^1(0,T; V')$ exist and we have $u'' + Au =: F \in L^1(0,T; H)$.*

*Then $u \in C([0,T]; V) \cap C^1([0,T]; H)$.*

*Proof.* The basic result goes back to [Str66] and we follow their ideas. Note though, that we have weaker assumptions on $A$ and $A'$. We also want to fix some small flaws in their proof.

Our main claim is that the energy

$$E(t) = \|u'(t)\|_H^2 + \langle A(t)u(t), u(t)\rangle_{V',V}$$

is continuous in $[0,T]$ even though $u'$ does not necessarily take values in $V$ as in Lemma A.1. For the moment, let us assume this claim is true and show that it proves the theorem.

Let $t_n \to t$. Clearly, $u \in C([0,T];H) \cap C^1([0,T];V')$. Therefore, $u(t_n) \to u(t)$ in $H$ and $u'(t_n) \to u'(t)$ in $V'$. Since we also have $u \in L^\infty(0,T;V)$ and $u' \in L^\infty(0,T;H)$ and since $V$ is reflexive, we get $u(t_n) \rightharpoonup u(t)$ in $V$ and $u'(t_n) \rightharpoonup u'(t)$ in $H$. The continuity of the energy now ensures that this weak convergence is actually strong convergence. In more detail,

$$\|u'(t) - u'(t_n)\|_H^2 + \langle A(t)(u(t) - u(t_n)), u(t) - u(t_n)\rangle_{V',V}$$
$$= E(t) + E(t_n) - 2(u'(t), u'(t_n))_H - 2\langle A(t)u(t), u(t_n)\rangle_{V',V}$$
$$\to 0.$$

Therefore, $u'(t_n) \to u'(t)$ in $H$ and, using the Gårding inequality, also $u(t_n) \to u(t)$ in $V$.

Now we come to the main part of the proof. We have to show that the energy is continuous in $[0,T]$. Actually, we will even show that $E$ is absolutely continuous with

$$E'(t) = \langle A'(t)u(t), u(t)\rangle_{V',V} + 2(F(t), u'(t))_H.$$

Fix $0 \le t_0 < t_1 \le T$, let $\delta, \varepsilon > 0$ with $2\delta < t_1 - t_0$ and define a continuous cutoff $\theta_\delta$ by setting $\theta_\delta(t) = 1$ for $t \in [t_0+\delta, t_1-\delta]$, $\theta_\delta(t) = 0$ for $t \notin [t_0, t_1]$ and affine on each of the two remaining intervals. Let $\eta$ be the standard smoothing kernel and as always $\eta_\varepsilon(t) = \varepsilon^{-1}\eta(\frac{t}{\varepsilon})$. Set $w_1 := \eta_\varepsilon * (\theta_\delta u') \in C_c^\infty(\mathbb{R};H)$ and $w_2 := \eta_\varepsilon * (\theta_\delta u) \in C_c^\infty(\mathbb{R};V)$. If we extend $A$ by $A(0)$ to the left and $A(T)$ to the right with $A'(t) = 0$ outside of $[0,T]$, we can use Lemma A.1 on some larger interval and get

$$0 = \int_{\mathbb{R}} 2(w_1(t), w_1'(t))_H + \langle A'(t)w_2(t), w_2(t)\rangle + 2\langle A(t)w_2(t), w_2'(t)\rangle \, dt.$$

That is,

$$0 = \int_{\mathbb{R}} 2(\eta_\varepsilon * (\theta_\delta u'), \eta_\varepsilon' * (\theta_\delta u'))_H + \langle A' \eta_\varepsilon * (\theta_\delta u), \eta_\varepsilon * (\theta_\delta u) \rangle_{V',V}$$
$$+ 2\langle A \eta_\varepsilon * (\theta_\delta u), \eta_\varepsilon' * (\theta_\delta u) \rangle_{V',V} \, dt.$$

Since

$$\eta_\varepsilon' * (\theta_\delta u') = \eta_\varepsilon * (\theta_\delta' u') + \eta_\varepsilon * (\theta_\delta u'')$$

pointwise in $V'$, we see that $\eta_\varepsilon * (\theta_\delta u'')$ actually takes values in $H$ and is even bounded in $L^\infty(\mathbb{R}; H)$ for fixed $\varepsilon$ and varying $\delta$, since $\theta_\delta$ and $\theta_\delta'$ are bounded in $L^1(\mathbb{R})$. Similarly,

$$\eta_\varepsilon' * (\theta_\delta u) = \eta_\varepsilon * (\theta_\delta' u) + \eta_\varepsilon * (\theta_\delta u')$$

and $\eta_\varepsilon * (\theta_\delta u')$ is bounded in $L^\infty(\mathbb{R}; V)$ for fixed $\varepsilon$ and varying $\delta$.

Thus, we can rewrite the above equality and get

$$0 = \int_{\mathbb{R}} 2(\eta_\varepsilon * (\theta_\delta u'), \eta_\varepsilon * (\theta_\delta' u'))_H + 2\langle \eta_\varepsilon * (\theta_\delta u''), \eta_\varepsilon * (\theta_\delta u') \rangle_{V',V}$$
$$+ \langle A'(\eta_\varepsilon * (\theta_\delta u)), \eta_\varepsilon * (\theta_\delta u) \rangle_{V',V}$$
$$+ 2\langle A(\eta_\varepsilon * (\theta_\delta u)) - \eta_\varepsilon * (\theta_\delta A u), \eta_\varepsilon' * (\theta_\delta u) \rangle_{V',V}$$
$$+ 2\langle \eta_\varepsilon * (\theta_\delta A u), \eta_\varepsilon * (\theta_\delta' u) \rangle_{V',V} + 2\langle \eta_\varepsilon * (\theta_\delta A u), \eta_\varepsilon * (\theta_\delta u') \rangle_{V',V} \, dt.$$

Now we want to let $\delta \to 0$. Since $\theta_\delta \to \chi_{[t_0,t_1]}$ in the $(L^\infty, L^1)$-Mackey topology, we have

$$\eta_\varepsilon * (\theta_\delta u) \to \eta_\varepsilon * (\chi_{[t_0,t_1]} u) \qquad \text{in } L^\infty(\mathbb{R}; V)$$
$$\eta_\varepsilon * (\theta_\delta u') \to \eta_\varepsilon * (\chi_{[t_0,t_1]} u') \qquad \text{in } L^\infty(\mathbb{R}; H)$$
$$\eta_\varepsilon * (\theta_\delta F) \to \eta_\varepsilon * (\chi_{[t_0,t_1]} F) \qquad \text{in } L^\infty(\mathbb{R}; H)$$
$$\eta_\varepsilon * (\theta_\delta A u) \to \eta_\varepsilon * (\chi_{[t_0,t_1]} A u) \qquad \text{in } L^\infty(\mathbb{R}; V')$$
$$\eta_\varepsilon' * (\theta_\delta u) \to \eta_\varepsilon' * (\chi_{[t_0,t_1]} u) \qquad \text{in } L^\infty(\mathbb{R}; V).$$

Therefore, we have

$$0 = \lim_{\delta \to 0} \int_{\mathbb{R}} 2(\eta_\varepsilon * \eta_\varepsilon * (\chi_{[t_0,t_1]} u'), \theta_\delta' u')_H$$
$$+ 2(\eta_\varepsilon * (\chi_{[t_0,t_1]} F), \eta_\varepsilon * (\chi_{[t_0,t_1]} u'))_H$$
$$+ \langle A'(\eta_\varepsilon * (\chi_{[t_0,t_1]} u)), \eta_\varepsilon * (\chi_{[t_0,t_1]} u) \rangle_{V',V}$$
$$+ 2\langle A(\eta_\varepsilon * (\chi_{[t_0,t_1]} u)) - \eta_\varepsilon * (\chi_{[t_0,t_1]} A u), \eta_\varepsilon' * (\chi_{[t_0,t_1]} u) \rangle_{V',V}$$
$$+ 2\langle \eta_\varepsilon * \eta_\varepsilon * (\chi_{[t_0,t_1]} A u), \theta_\delta' u) \rangle_{V',V} \, dt.$$

The maps

$$t \mapsto (\eta_\varepsilon * \eta_\varepsilon * (\chi_{[t_0,t_1]} u')(t), u'(t))_H$$

and

$$t \mapsto \langle \eta_\varepsilon * \eta_\varepsilon * (\chi_{[t_0,t_1]} Au)(t), u(t) \rangle_{V',V}$$

are continuous in $[t_0, t_1]$ as a product of a continuous and a weakly continuous function and as a product of a continuous and a weak-$*$ continuous function, respectively. Plugging in the actual values of $\theta_\delta'$, we therefore get

$$\begin{aligned} 0 = {} & 2(\eta_\varepsilon * \eta_\varepsilon * (\chi_{[t_0,t_1]} u')(t_0), u'(t_0))_H - 2(\eta_\varepsilon * \eta_\varepsilon * (\chi_{[t_0,t_1]} u')(t_1), \\ & u'(t_1))_H + 2\langle \eta_\varepsilon * \eta_\varepsilon * (\chi_{[t_0,t_1]} Au)(t_0), u(t_0) \rangle_{V',V} \\ & - 2\langle \eta_\varepsilon * \eta_\varepsilon * (\chi_{[t_0,t_1]} Au)(t_1), u(t_1) \rangle_{V',V} \\ & + \int_{\mathbb{R}} 2(\eta_\varepsilon * (\chi_{[t_0,t_1]} F), \eta_\varepsilon * (\chi_{[t_0,t_1]} u'))_H \\ & + \langle A'(\eta_\varepsilon * (\chi_{[t_0,t_1]} u)), \eta_\varepsilon * (\chi_{[t_0,t_1]} u) \rangle_{V',V} \\ & + 2\langle A(\eta_\varepsilon * (\chi_{[t_0,t_1]} u)) - \eta_\varepsilon * (\chi_{[t_0,t_1]} Au), \eta_\varepsilon' * (\chi_{[t_0,t_1]} u) \rangle_{V',V} \, dt. \end{aligned}$$

Next, we want to send $\varepsilon \to 0$. For any $v \in C_c^1(\mathbb{R}; V)$, we have $\varepsilon \eta_\varepsilon' * v \to 0$ in $L^2(\mathbb{R}; V)$. Therefore, for any such $v$,

$$\limsup_{\varepsilon \to 0} \| \varepsilon \eta_\varepsilon' * (\chi_{[t_0,t_1]} u) \|_{L^2(\mathbb{R};V)} \leq \| \eta' \|_{L^1(\mathbb{R})} \| \chi_{[t_0,t_1]} u - v \|_{L^2(\mathbb{R};V)}.$$

Hence, $\varepsilon \eta_\varepsilon' * (\chi_{[t_0,t_1]} u) \to 0$ in $L^2(\mathbb{R}; V)$. At the same time

$$\begin{aligned} \| A(\eta_\varepsilon * (\chi_{[t_0,t_1]} u)) & - \eta_\varepsilon * (\chi_{[t_0,t_1]} Au) \|_{L^2(\mathbb{R};V')} \\ & \leq \varepsilon \| A' \|_{L^\infty} \| \eta_\varepsilon * (\chi_{[t_0,t_1]} u) \|_{L^2(\mathbb{R};V')} \end{aligned}$$

and so the last term in the integral goes to 0.

Now $\eta_\varepsilon * \eta_\varepsilon \geq 0$ with $\int_0^\infty \eta_\varepsilon * \eta_\varepsilon \, dt = \frac{1}{2}$. Therefore, by Lebesgue's theorem and the weak continuity of $u'$ in $H$,

$$\begin{aligned} (\eta_\varepsilon * \eta_\varepsilon * (\chi_{[t_0,t_1]} u')(t_0), u'(t_0))_H &= \int_0^\infty (u'(t_0 + \varepsilon s), u'(t_0))_H \, \eta * \eta(s) \, ds \\ &\to \int_0^\infty (u'(t_0), u'(t_0))_H \, \eta * \eta(s) \, ds \\ &= \frac{1}{2}(u'(t_0), u'(t_0))_H. \end{aligned}$$

Similar results hold for the other terms and we conclude that

$$E(t_1) - E(t_0) = \int_{t_0}^{t_1} 2(F, u')_H + \langle A'u, u \rangle_{V',V} \, dt$$

for any $0 \leq t_0 < t_1 \leq T$.                                              □

# Appendix B

# Multiplication and Composition of Sobolev Functions

The following Lemma is very useful to control products of Sobolev functions with the same integrability exponent $p$.

**Lemma B.1.** *Let $1 \leq p < \infty$ and $K, M \in \mathbb{N}$, such that $K \geq 1$ and $M > \frac{d}{p}$ and let $\lambda_k \in \mathbb{N}_0$ for $1 \leq k \leq K$ with $\sum_{k=1}^{K} \lambda_k =: N \leq M$. Let $\Omega \subset \mathbb{R}^d$ be open and bounded with $\partial\Omega$ Lipschitz. Then there is a $C > 0$ such that for any $f_k \in W^{M-\lambda_k,p}$ we have $\prod_{k=1}^{K} f_k \in W^{M-N,p}(\Omega)$ with*

$$\| \prod_{k=1}^{K} f_k \|_{W^{M-N,p}(\Omega)} \leq C \prod_{k=1}^{K} \|f_k\|_{W^{M-\lambda_k,p}(\Omega)}.$$

*Additionally, the product mapping is continuous even from the weak topologies on the $W^{M-\lambda_k,p}(\Omega)$ to the strong topology on $W^{L,p}(\Omega)$ if either $L < M - N$ or $L = M - N$ and $\lambda_k < N$ for all $k$.*

*In particular, $W^{M,p}(\Omega)$ is a Banach algebra.*

*Proof.* By density, it suffices to consider functions in $C^\infty(\overline{\Omega})$. Furthermore, by the product rule, it suffices to prove $\prod_{k=1}^{K} f_k \in L^p(\Omega)$ with

$$\| \prod_{k=1}^{K} f_k \|_{L^p(\Omega)} \leq C \prod_{k=1}^{K} \|f_k\|_{W^{M-\lambda_k,p}(\Omega)}.$$

Standard embedding theorems give $W^{M-l,p} \hookrightarrow L^\infty$ for $M - l > \frac{d}{p}$, $W^{M-\lambda,p} \hookrightarrow L^q$ for any $1 \leq q < \infty$ and $M - l = \frac{d}{p}$ and $W^{M-l,p} \hookrightarrow L^q$

for $M - l < \frac{d}{p}$ and $q \leq \frac{d}{\frac{d}{p} - (M-l)}$. Let us write

$$A = \{k \colon \lambda_k > M - \frac{d}{p}\}$$

$$B = \{k \colon \lambda_k = M - \frac{d}{p}\}.$$

Hölder's inequality now directly establishes the claim if either $B = \emptyset$ and $\sum_{k \in A} \frac{\frac{d}{p} - M + \lambda_k}{\frac{d}{p}} \leq 1$ or $B \neq \emptyset$ and $\sum_{k \in A} \frac{\frac{d}{p} - M + \lambda_k}{\frac{d}{p}} < 1$. This is trivially true if $A = \emptyset$ or if $|A| = 1$ and $B = \emptyset$. If now $|A| = 1$ but $B \neq \emptyset$, writing $A = \{k_0\}$, we find that $\lambda_{k_0} \leq N - (M - \frac{d}{2}) < N \leq M$ and the condition is satisfied. Finally, if $|A| \geq 2$ then

$$\sum_{k \in A} \frac{\frac{d}{2} - M + \lambda_k}{\frac{d}{2}} \leq \frac{|A|(\frac{d}{2} - M) + N}{\frac{d}{2}}$$

$$\leq \frac{2(\frac{d}{2} - M) + M}{\frac{d}{2}}$$

$$= 1 - \frac{M - \frac{d}{2}}{\frac{d}{2}} < 1.$$

For the additional claim, note that, even after taking up to $L$ derivatives with the product rule, we always have $\lambda_k < M$, so that all the functions are embedded into some space of lower differentiability. The embeddings were already compact if $M - \lambda_k \geq \frac{d}{p}$. To control the other case we just have to make sure that we always have the strict inequality $\sum_{k \in A} \frac{\frac{d}{p} - M + \lambda_k}{\frac{d}{p}} < 1$. This was only unclear in the case $|A| = 1$, $B = \emptyset$. But since $\lambda_k < M$, we now have a strict inequality in this case too.  $\square$

**Lemma B.2.** *Given $f, g \in W^{n,p}(\Omega)$ for some open and bounded set $\Omega \subset \mathbb{R}^d$ with Lipschitz boundary. If $np > d$, then $f \cdot g \in W^{n,p}(\Omega)$ and*

$$\|fg\|_{W^{n,p}(\Omega)} \leq C \|f\|_{W^{n,p}(\Omega)} \|g\|_{W^{n,p}(\Omega)}$$

*for some $C = C(n, p, \Omega) > 0$.*

*The cone condition is satisfied for instance if $\Omega$ is bounded and has a Lipschitz boundary.*

*Proof.* This is a well known result. It can be found, e.g., in [AF03, Thm. 4.39]. It also follows directly from Lemma B.1, by setting $M = n$, $K = 2$, and $\lambda_1 = \lambda_2 = 0$.  $\square$

We also need the following multivariate version of the Faà di Bruno formula:

**Lemma B.3.** *Let* $n, d, k, l \in \mathbb{N}$. *Let* $\Omega \subset \mathbb{R}^d$ *open,* $g \in C^n(\Omega; \mathbb{R}^k)$ *and* $f \in C^n(\mathbb{R}^k; \mathbb{R}^l)$. *Then* $f \circ g \in C^n(\Omega; \mathbb{R}^l)$ *with*

$$D^\alpha(f \circ g)(x) = \sum_{\substack{\beta \in \mathbb{N}_0^k \\ 1 \le |\beta| \le |\alpha|}} D^\beta f(g(x)) \sum_{s=1}^{|\alpha|} \sum_{p_s(\alpha, \beta)} \alpha! \prod_{j=1}^s \frac{(D^{\gamma_j} g(x))^{\lambda_j}}{\lambda_j! (\gamma_j!)^{|\lambda_j|}}$$

*for all* $x \in \Omega$ *and* $\alpha \in \mathbb{N}_0^d$ *with* $1 \le |\alpha| \le n$, *where*

$$p_s(\alpha, \beta) = \Big\{ (\lambda_1, \ldots, \lambda_s; \gamma_1, \ldots, \gamma_s) \colon \lambda_j \in \mathbb{N}_0^k, \gamma_j \in \mathbb{N}_0^d,$$

$$0 \prec \gamma_1 \prec \cdots \prec \gamma_s, |\lambda_j| > 0, \sum_{j=1}^s \lambda_j = \beta, \sum_{j=1}^s \gamma_j |\lambda_j| = \alpha \Big\}$$

*and* $\gamma_1 \prec \gamma_2$ *if and only if* $|\gamma_1| < |\gamma_2|$ *or* $|\gamma_1| = |\gamma_2|$ *and, for some* $j$,

$$(\gamma_1)_1 = (\gamma_2)_1, \ldots, (\gamma_1)_{j-1} = (\gamma_2)_{j-1} \text{ and } (\gamma_1)_j < (\gamma_2)_j.$$

*Proof.* Even though it is quite possible that the result itself is much older, it can be found in [CS96]. $\qquad\square$

While the result is only stated for strongly differentiable functions, one can always extend it to the case where $g$ and then $f \circ g$ are Sobolev functions if one can estimate the right hand side suitably.

As a corollary we also get the following statements.

**Corollary B.4.** *Let* $n, d, k, l \in \mathbb{N}$. *There are* $C = C(n, d, k, l) > 0$ *such that the following holds. Let* $\Omega \subset \mathbb{R}^d$ *open,* $g \in C^n(\Omega; \mathbb{R}^k)$ *and* $f \in C^n(\mathbb{R}^k; \mathbb{R}^l)$. *Then*

$$|D^n(f \circ g)(x)| \le C \sum_{s=1}^n |D^s f(g(x))| \sum_{\substack{l_1, \ldots, l_s \ge 1 \\ l_1 + \cdots + l_s = n}} \prod_{j=1}^s |D^{(l_j)} g(x)|,$$

*for all* $x \in \Omega$. *Furthermore, if* $f$ *and all its derivatives are uniformly continuous and* $h \in C^n(\Omega; \mathbb{R}^k)$ *then*

$$|D^n(f \circ (g+h))(x) - D^n(f \circ g)(x)|$$

$$\leq C \sum_{s=1}^{n} \sum_{\substack{l_1,\ldots,l_s \geq 1 \\ l_1+\cdots+l_s=n}} \omega_{D^s f}(|h(x)|) \prod_{j=1}^{s} |D^{l_j}g(x)|$$

$$+ C \sum_{s=1}^{n} \sum_{\substack{l_1,\ldots,l_s \geq 1 \\ l_1+\cdots+l_s=n}} \sum_{m=1}^{s} |D^s f(g(x)+h(x))| \prod_{j=1}^{m} |D^{l_j}h(x)| \prod_{j=m+1}^{s} |D^{l_j}g(x)|$$

*Proof.* The first estimate follows directly from Lemma B.3. For the second estimate one can also use the lemma and then calculate

$$\left| D^\beta f(g(x)+h(x)) \prod_{j=1}^{s} (D^{\gamma_j}(g+h)(x))^{\lambda_j} - D^\beta f(g(x)) \prod_{j=1}^{s} (D^{\gamma_j}g(x))^{\lambda_j} \right|$$

$$\leq \omega_{D^\beta f}(|h(x)|) \prod_{j=1}^{s} |D^{|\gamma_j|}g(x)|^{|\lambda_j|}$$

$$+ \left| D^\beta f(g(x)+h(x)) \right| \left| \prod_{j=1}^{s} (D^{\gamma_j}(g+h)(x))^{\lambda_j} - \prod_{j=1}^{s} (D^{\gamma_j}g(x))^{\lambda_j} \right|$$

$$\square$$

**Lemma B.5.** *Let* $m \in \mathbb{N}_0$, $d < 2m+2$ *and let* $\Omega \subset \mathbb{R}^d$ *be an open, bounded set with Lipschitz boundary. Let* $V \subset \mathbb{R}^{d \times \mathcal{R}}$ *be open and* $W_{\mathrm{atom}} \in C^{m+2}(V)$.

*Now define the operator* $F: B \mapsto DW_{\mathrm{CB}} \circ B$. *Then*

$$\{B \in H^{m+1}(\Omega; \mathbb{R}^{d \times d}): \inf_{x \in \Omega} \mathrm{dist}((B(x)\rho)_{\rho \in \mathcal{R}}, V^c) > 0\}$$

*is open in* $H^{m+1}(\Omega; \mathbb{R}^{d \times d})$ *and*

$$F: \{B \in H^{m+1}(\Omega; \mathbb{R}^{d \times d}): \inf_{x \in \Omega} \mathrm{dist}((B(x)\rho)_{\rho \in \mathcal{R}}, V^c) > 0\}$$
$$\to H^{m+1}(\Omega; \mathbb{R}^{d \times d})$$

*is well-defined, continuous and bounded. Furthermore, if even* $W_{\mathrm{atom}} \in C^{m+3}(V)$, *then* $F$ *is* $C^1$ *with*

$$DF(B)[H](x) = D^2 W_{\mathrm{CB}}(B(x))[H(x)].$$

*Proof.* This is not new and for the most part contained, e.g., in [Val88, I. Thm.3.1] and [Val88, II. Thm.4.1]. Only the boundedness is not explicitly mentioned, but it follows along the same lines. □

# Appendix C

# Elliptic Regularity for Sobolev Coefficients

We need a result on higher order regularity for linear systems that are elliptic in the Legendre-Hadamard sense. To be useful for quasilinear equations it is crucial that the regularity assumptions on the coefficients are not too strong. In the standard literature the typical assumption for $W^{k+2,p}$-regularity of the solution is $A \in C^{k,1} = W^{k+1,\infty}$ or, more rarely, $A \in W^{k+1,p}$ if $p > d$. We will reduce the last assumption to the much weaker condition $p(k+1) > d$. It is no coincidence that this assumption corresponds to what is needed for $A$ to be continuous. Actually, this is known to be the critical case. It seems reasonable that the case $p(k+1) = d$ can be included, as there are regularity results where the coefficients are not continuous but only have vanishing mean oscillation, but we will not investigate this question here.

For the sake of generality we will consider the general case $1 < p < \infty$ but we are mostly interested in the case $p = 2$. Even though it seems quite possible that this kind of result has been proven before, it does not seem to be available in the standard literature. It is largely, but not quite, contained in [SS09] and, of course, builds heavily on the famous classical work [ADN64].

We consider a differential operator in divergence form

$$(Lu)_i = (-\operatorname{div}(A\nabla u) + b\nabla u)_i = -\sum_{j,k,l} \frac{\partial}{\partial x_j}\left(A_{ijkl}\frac{\partial u_k}{\partial x_l}\right) + \sum_{k,l} b_{ikl}\frac{\partial u_k}{\partial x_l}.$$

In particular, we are interested in the cases $b = 0$ and, if $A$ is Lipschitz, $b_{ikl} = \sum_j \frac{\partial A_{ijkl}}{\partial x_j}$. The second case corresponds to an operator in non-divergence form

$$(Lu)_i = \sum_{j,k,l} A_{ijkl} \frac{\partial^2 u_k}{\partial x_j \partial x_l}.$$

On some open set $\Omega \subset \mathbb{R}^d$ define the corresponding bilinear form

$$B(u,v) = \int_\Omega \sum_{i,j,k,l} A_{ijkl} \frac{\partial v_i}{\partial x_j} \frac{\partial u_k}{\partial x_l} + \sum_{i,k,l} b_{ikl} v_i \frac{\partial u_k}{\partial x_l} \, dx.$$

whenever it is well-defined.

Let us first recall the classical Gårding inequality:

**Theorem C.1.** *Let $\Omega \subset \mathbb{R}^d$ be an open and bounded set and $\lambda_0 > 0$ such that*

$$\sum_{i,j,k,l} A_{ijkl}(x) \xi_i \eta_j \xi_k \eta_l \geq \lambda_0 |\xi|^2 |\eta|^2$$

*for all $x \in \Omega, \xi \in \mathbb{R}^N, \eta \in \mathbb{R}^d$. Furthermore, assume that $A$ is bounded and uniformly continuous with modulus $\omega$ and $b \in L^\infty(\Omega)$.*

*Then there exists a $\lambda_1 = \lambda_1(\|A\|_\infty, \lambda_0, \Omega, \omega, \|b\|_\infty) \geq 0$, such that*

$$\frac{\lambda_0}{2} \int_\Omega |\nabla u|^2 \, dx \leq B(u,u) + \lambda_1 \int_\Omega |u|^2 \, dx$$

*for all $u \in H_0^1(\Omega; \mathbb{R}^N)$.*

*If $A$ is constant and $b = 0$, we can take $\lambda_1 = 0$ and can even achieve $\lambda_0$ instead of $\frac{\lambda_0}{2}$ as the constant on the left side.*

We have the following a priori estimates:

**Theorem C.2.** *Let $k \in \mathbb{N}_0$ and let $\Omega \subset \mathbb{R}^d$ be open and bounded with $C^{k+2}$ boundary. Let $1 < p < \infty$ and assume that $(k+1)p > d$. Let $\lambda, \Lambda > 0$ and let $A \in W^{k+1,p}(\Omega; \mathbb{R}^{d \times d \times d \times d})$, such that $\|A\|_{W^{k+1,p}(\Omega)} \leq \Lambda$ and*

$$A(x)[\xi \otimes \eta, \xi \otimes \eta] \geq \lambda |\xi|^2 |\eta|^2$$

*for all $\xi, \eta \in \mathbb{R}^d$ and $x \in \Omega$. Then there is a $C = C(\Omega, \lambda, \Lambda, p, k)$ such that for all $r \in \{0, \ldots, k\}$ and $u \in W^{r+2,p}(\Omega; \mathbb{R}^d)$ we have*

$$\|u\|_{W^{r+2,p}(\Omega)} \leq C(\|\mathrm{div}(A\nabla u)\|_{W^{r,p}(\Omega)} + \|u\|_{L^p(\Omega)}).$$

*We also have the estimate in non-divergence form*

$$\|u\|_{W^{r+2,p}(\Omega)} \leq C\Big(\Big\|\Big(\sum_{j,k,l} A_{ijkl} \frac{\partial^2 u_k}{\partial x_j \partial x_l}\Big)_i\Big\|_{W^{r,p}(\Omega)} + \|u\|_{L^p(\Omega)}\Big).$$

*Proof.* In [SS09] this has been proven in the case $r = k$ in divergence form based on the estimates of [ADN64] for constant coefficients. But all other cases follow mostly along the same lines. This includes the case in non-divergence form since the proof is based on approximating $A$ locally by a constant. The case of smaller $r$ follows along the same lines as well, since the Agmon-Douglis-Nirenberg estimates are still valid. The only difference in all these cases is that one can no longer use the tame estimate of Moser. Instead one has to use the following finer estimates for multiplications in Sobolev spaces, namely

$$\|J\partial_l u\|_{W^{r+1,p}(\Omega)} \leq \varepsilon \|J\|_{W^{k+1,p}(\Omega)}\|u\|_{W^{r+2,p}(\Omega)} + C_\varepsilon \|J\|_{W^{k+1,p}(\Omega)}\|u\|_{L^p(\Omega)} + C\|J\|_{L^\infty(\Omega)}\|u\|_{W^{r+2,p}(\Omega)}$$

and

$$\|J\partial_j\partial_l u\|_{W^{r,p}(\Omega)} \leq \varepsilon \|J\|_{W^{k+1,p}(\Omega)}\|u\|_{W^{r+2,p}(\Omega)} + C_\varepsilon \|J\|_{W^{k+1,p}(\Omega)}\|u\|_{L^p(\Omega)} + C\|J\|_{L^\infty(\Omega)}\|u\|_{W^{r+2,p}(\Omega)}$$

for all $J \in W^{k+1,p}(\Omega), u \in W^{r+2,p}(\Omega)$, $\varepsilon > 0$, $0 \leq r \leq k$, $j$ and $l$ with constants that may depend on $p$, $k$ and $\Omega$. Both inequalities can be proven rather easily using the product rule, the Sobolev and Rellich–Kondrachov embedding theorems, as well as Ehrling's lemma. $\square$

**Theorem C.3.** *Let $k \in \mathbb{N}_0$ and let $\Omega \subset \mathbb{R}^d$ be open and bounded with $C^{k+2}$ boundary. Let $1 < p < \infty$, $r \in \{0,\ldots,k\}$ and assume that $(k+1)p > d$ and $p \geq \frac{2d}{d+2(r+1)}$. Let $\lambda, \Lambda_1, \Lambda_2 > 0$ and let $A \in W^{k+1,p}(\Omega; \mathbb{R}^{d\times d\times d\times d})$, such that $\|A\|_{W^{k+1,p}(\Omega)} \leq \Lambda_1$ and*

$$A(x)[\xi \otimes \eta, \xi \otimes \eta] \geq \lambda |\xi|^2 |\eta|^2$$

*for all $\xi, \eta \in \mathbb{R}^d$ and $x \in \Omega$. Consider*

$$L_{1,\mu}, L_{2,\mu} \colon W^{r+2,p}(\Omega; \mathbb{R}^d) \cap H_0^1(\Omega; \mathbb{R}^d) \to W^{r,p}(\Omega; \mathbb{R}^d)$$

*defined by*

$$L_{1,\mu}u = -\operatorname{div}(A\nabla u) + \mu u,$$
$$(L_{2,\mu}u)_i = -\sum_{j,k,l} A_{ijkl}\frac{\partial^2 u_k}{\partial x_j \partial x_l} + \mu u_i.$$

*There is a $\mu_1 = \mu_1(\Omega, k, p, \Lambda_1, \lambda)$ such that for all $\mu \geq \mu_1$ $L_{1,\mu}$ is an isomorphism. If additionally we have $A \in W^{1,\infty}(\Omega; \mathbb{R}^{d \times d \times d \times d})$ with $\|A\|_{W^{1,\infty}} \leq \Lambda_2$, then there is a $\mu_2 = \mu_2(\Omega, k, p, \Lambda_1, \Lambda_2, \lambda)$ such that for all $\mu \geq \mu_2$ $L_{2,\mu}$ is an isomorphism. Furthermore, we have the estimates*

$$\|L_{i,\mu}^{-1}\| \leq C_i,$$

*where $C_1 = C_1(\Omega, \lambda, \Lambda_1, p, k, \mu_{\max}) > 0$ and $C_2 = C_2(\Omega, \lambda, \Lambda_1, \Lambda_2, p, k, \mu_{\max}) > 0$ and $\mu_i \leq \mu \leq \mu_{\max}$.*

*Proof.* We argue by continuity. Set $(A^t)_{ijkl} = t A_{ijkl} + (1 - t)\delta_{ik}\delta_{jl}$ for $t \in [0, 1]$ and denote by $L_{1,\mu}^t, L_{2,\mu}^t$ the corresponding operators. Since $(k + 1)p > d$, these operators are well defined by Lemma B.1 with

$$\|L_{i,\mu}^t\|_{L(W^{r+2,p}(\Omega), W^{r,p}(\Omega))} \leq 1 + |\mu| + C(\Omega, k, p)\|A\|_{W^{k+1,p}(\Omega)}$$

for all $t, i, \mu$.

   *Claim 1: There are $\mu_i$, $i = 1, 2$, such that $L_{i,\mu}^t$ is one-to-one for all $\mu \geq \mu_i$.*

We can apply Theorem C.1. Note that the modulus $\omega$ can be chosen only dependent on $\Omega, k, p, \Lambda$. If $L_{1,\mu}^t u = 0$, we can apply Theorem C.1 with $b = 0$ to obtain

$$\frac{\lambda}{2} \int_\Omega |\nabla u|^2 \, dx \leq 0$$

and thus $u = 0$, whenever $\mu \geq \mu_1 = \mu_1(\Omega, k, p, \Lambda_1, \lambda)$. If $L_{2,\mu}^t u = 0$, we can apply Theorem C.1 with $b_{ikl} = -\sum_j \frac{\partial A_{ijkl}}{\partial x_j}$ to obtain

$$\frac{\lambda}{2} \int_\Omega |\nabla u|^2 \, dx \leq 0$$

and thus $u = 0$, whenever $\mu \geq \mu_2 = \mu_2(\Omega, k, p, \Lambda_1, \Lambda_2, \lambda)$.

   *Claim 2: There are constants $C_1, C_2 > 0$ with $C_1 = C_1(\Omega, \lambda, \Lambda_1, p, k, \mu_{\max})$ and $C_2 = C_2(\Omega, \lambda, \Lambda_1, \Lambda_2, p, k, \mu_{\max})$ such that for all $t \in [0, 1]$, $u \in W^{r+2,p}(\Omega; \mathbb{R}^d) \cap H_0^1(\Omega; \mathbb{R}^d)$, $i \in \{1, 2\}$ and $\mu_i \leq \mu \leq \mu_{\max}$ we have*

$$\|u\|_{W^{r+2,p}(\Omega)} \leq C_i \|L_{i,\mu}^t u\|_{W^{r,p}(\Omega)}.$$

We argue by contradiction. If there were no such $C$, then there exist $t_n, \mu_n, u_n, A_n$ such that

$$1 = \|u_n\|_{W^{r+2,p}(\Omega)} > n\|L_{i,\mu_n}^{t_n}(A_n)u_n\|_{W^{r,p}(\Omega)}.$$

Furthermore,

$$A_n(x)[\xi \otimes \eta, \xi \otimes \eta] \geq \lambda |\xi|^2 |\eta|^2$$

for all $\xi, \eta \in \mathbb{R}^d$ and $x \in \Omega$ and $\|A_n\|_{W^{k+1,p}} \leq \Lambda_1$. Then we find a subsequence (not relabeled), such that $t_n \to t$, $\mu_n \to \mu$ and $u_n \rightharpoonup u$ in $W^{r+2,p}$ and $A_n \rightharpoonup A$ in $W^{k+1,p}$. In particular $A_n \to A$ uniformly, $A$ is still elliptic with constant $\lambda$ and $u_n \to u$ strongly in $W^{r+1,p}$ and weakly in $H_0^1$. If $i = 1$ we easily deduce that $L_{1,\mu_n}^{t_n}(A_n)u_n \to L_{1,\mu}^t(A)u$ in distribution. If $i = 2$ we additional have $\|A_n\|_{W^{1,\infty}} \leq \Lambda_2$. By uniform convergence we also have $\|A\|_{W^{1,\infty}} \leq \Lambda_2$. We also have directly $L_{2,\mu_n}^{t_n}(A_n)u_n \rightharpoonup L_{2,\mu}^t(A)u$ in $L^p$. But in both cases we also now that $\|L_{i,\mu_n}^{t_n}(A_n)u_n\|_{W^{r,p}(\Omega)} \to 0$. Hence, $L_{i,\mu}^t(A)u = 0$ with $u \in W^{r+2,p}(\Omega;\mathbb{R}^d) \cap H_0^1(\Omega;\mathbb{R}^d)$ and thus $u = 0$ by claim 1. Now we use Theorem C.2 to find

$$\begin{aligned}
1 &= \|u_n\|_{W^{r+2,p}(\Omega)} \\
&\leq C\Big(\|L_{i,\mu_n}^{t_n}(A_n)u_n\|_{W^{r,p}(\Omega)} + |\mu_n|\|u_n\|_{W^{r,p}(\Omega)} + \|u_n\|_{L^p(\Omega)}\Big)
\end{aligned}$$

Since the right hand side goes to 0, we have a contradiction.

*Claim 3: For $\mu \geq \mu_i$ the sets*

$$I_{i,\mu} = \{t \in [0,1] \colon L_{i,\mu}^t \text{ is onto}\}$$

*are closed in $[0,1]$.*

Given $t_n \to t$, $t_n \in I_{i,\mu}$ and $f \in W^{r,p}(\Omega;\mathbb{R}^d)$, there are $u_n \in W^{r+2,p}(\Omega;\mathbb{R}^d) \cap H_0^1(\Omega;\mathbb{R}^d)$ such that $L_{i,\mu}^{t_n}u_n = f$. By claim 2 the $u_n$ are bounded in $W^{r+2,p}(\Omega;\mathbb{R}^d)$. On a subsequence (not relabeled) we thus find $u_n \rightharpoonup u$ in $W^{r+2,p}(\Omega;\mathbb{R}^d)$ and easily deduce $L_{i,\mu}^{t_n}u_n \to L_{i,\mu}^t u$ in the sense of distributions. Hence, $L_{i,\mu}^t u = f$ and $t \in I_{i,\mu}$.

*Claim 4: For $\mu \geq \mu_i$ the sets*

$$I_{i,\mu} = \{t \in [0,1] \colon L_{i,\mu}^t \text{ is onto}\}$$

*are open in $[0,1]$.*

Let $t \in I_{i,\mu}$. Then $L_{i,\mu}^t$ is continuous, onto and one-to-one and therefore an isomorphism by the closed graph theorem. Set

$$\delta = \frac{1}{2\|(L_{i,\mu}^t)^{-1}\|(\|L_{i,\mu}^1\| + \|L_{i,\mu}^0\|)}$$

and let $s \in [0,1]$ with $|s - t| < \delta$. Let $f \in W^{r,p}(\Omega;\mathbb{R}^d)$ and let $u_0 = (L_{i,\mu}^t)^{-1}f$. We have to find a $u \in W^{r+2,p}(\Omega;\mathbb{R}^d) \cap H_0^1(\Omega;\mathbb{R}^d)$

with $L_{i,\mu}^s u = f$ which is equivalent to finding a fixed point of

$$G_s(u) = (L_{i,\mu}^t)^{-1}(L_{i,\mu}^t u - L_{i,\mu}^s u + f).$$

We claim that $G_s \colon \overline{B_r(u_0)} \to \overline{B_r(u_0)}$ is well defined and a contraction for $r = \|u_0\|$. Indeed, since

$$\|L_{i,\mu}^t - L_{i,\mu}^s\| \le \delta(\|L_{i,\mu}^1\| + \|L_{i,\mu}^0\|),$$

we find

$$\begin{aligned}
\|G_s(u) - u_0\| &\le \|(L_{i,\mu}^t)^{-1}\|\|L_{i,\mu}^t u - L_{i,\mu}^s u\| \\
&\le 2r\delta\|(L_{i,\mu}^t)^{-1}\|(\|L_{i,\mu}^1\| + \|L_{i,\mu}^0\|) \\
&\le r
\end{aligned}$$

and

$$\|G_s(u) - G_s(v)\| \le \|(L_{i,\mu}^t)^{-1}\|\|L_{i,\mu}^t - L_{i,\mu}^s\|\|u - v\| \le \frac{1}{2}\|u - v\|.$$

Banach's fixed point theorem gives the desired result.

*Claim 5:* For $\mu \ge \mu_i$, we have $0 \in I_{i,\mu}$.

This is just the (scalar) Laplacian in each component. This is a well known result. E.g., this is a special case of results in [GT01].

Since $[0,1]$ is connected, we have shown that $I_{i,\mu} = [0,1]$ for $\mu \ge \mu_i$. In particular, $L_{1,\mu}$ and $L_{2,\mu}$ are isomorphisms. The estimates follow from claim 2. $\hfill\square$

# Bibliography

[AC04]    Roberto Alicandro and Marco Cicalese. A general integral representation result for continuum limits of discrete energies with superlinear growth. *SIAM J. Math. Anal.*, 36(1):1–37, 2004. doi:10.1137/S0036141003426471.

[ACG11]   Roberto Alicandro, Marco Cicalese, and Antoine Gloria. Integral representation results for energies defined on stochastic lattices and application to nonlinear elasticity. *Arch. Rational Mech. Anal.*, 200(3):881–943, 2011. doi:10.1007/s00205-010-0378-7.

[ADN64]   Shmuel Agmon, Avron Douglis, and Louis Nirenberg. Estimates near the boundary for solutions of elliptic partial differential equations satisfying general boundary conditions. II. *Comm. Pure Appl. Math.*, 17:35–92, 1964. doi:10.1002/cpa.3160170104.

[AF03]    Robert Adams and John J.F. Fournier. *Sobolev Spaces*. Elsevier Science Ltd, second edition, 2003.

[AFG00]   Roberto Alicandro, Matteo Focardi, and Maria Stella Gelli. Finite-difference approximation of energies in fracture mechanics. *Ann. Scuola Norm. Sup. Pisa Cl. Sci.*, 29(3):671–709, 2000. URL: http://www.numdam.org/item?id=ASNSP_2000_4_29_3_671_0.

[BD98]    Andrea Braides and Anneliese Defranceschi. *Homogenization of Multiple Integrals*. Oxford University Press, 1998.

[BLL02]   Xavier Blanc, Claude LeBris, and Pierre-Louis Lions. From molecular models to continuum mechanics. *Arch. Rational Mech. Anal.*, 164:341–381, 2002. doi:10.1007/s00205-002-0218-5.

[Bra85]      Andrea   Braides.      Homogenization   of  some  al-
             most  periodic  coercive  functional.     *Rend.  Ac-
             cad.  Naz.  Sci.  XL*,  IX(103):313–322,  1985.   URL:
             `http://www.accademiaxl.it/en/publications/`
             `academy-proceedings-on-line.html`.

[Bra16]      Julian Braun. Connecting atomistic and continuous models
             of elastodynamics. 2016. URL: `http://www.arxiv.org/`
             `abs/1606.01723`. preprint.

[BS13]       Julian Braun and Bernd Schmidt.    On the passage
             from atomistic systems to nonlinear elasticity theory
             for general multi-body potentials with p-growth.  *Net-
             works  and  Heterogeneous  Media*,  8(4):879–912,  2013.
             doi:10.3934/nhm.2013.8.879.

[BS16]       Julian Braun and Bernd Schmidt. Existence and conver-
             gence of solutions of the boundary value problem in atom-
             istic and continuum nonlinear elasticity theory. 2016. URL:
             `http://www.arxiv.org/abs/1604.00197`. preprint.

[CDKM06]     Sergio Conti, Georg Dolzmann, Bernd Kirchheim, and Ste-
             fan Müller.   Sufficient conditions for the validity of the
             Cauchy-Born rule close to $SO(n)$.   *J. Eur. Math. Soc.
             (JEMS)*, 8:515–539, 2006. doi:10.4171/JEMS/65.

[CS96]       Gregory M. Constantine and Thomas H. Savits. A multivari-
             ate Faa di Bruno formula with applications. *Trans. Amer.
             Math. Soc.*, 348(2):503–520, 1996. doi:10.1090/S0002-9947-
             96-01501-2.

[Dac08]      Bernard Dacorogna. *Direct Methods in the Calculus of Vari-
             ation*. Springer, 2008. doi:10.1007/978-0-387-55249-1.

[Dei10]      Klaus Deimling.  *Nonlinear Functional Analysis*.  Dover,
             2010.

[DH85]       Constantine M. Dafermos and William J. Hrusa.  Energy
             methods for quasilinear hyperbolic initial-boundary value
             problems. applications to elastodynamics. *Arch. Rational
             Mech. Anal.*, 87(3):267–292, 1985. doi:10.1007/BF00250727.

[DM93]      Gianni Dal Maso.  *An Introduction to* Γ*-Convergence.*
            Birkhäuser, 1993. doi:10.1007/978-1-4612-0327-8.

[EM07a]     Weinan E and Pingbing Ming. Cauchy-Born rule and the
            stability of crystalline solids: Dynamic problems. *Acta
            Mathematicae Applicatae Sinica, English Series*, 23:529–
            550, 2007. doi:10.1007/s10255-007-0393.

[EM07b]     Weinan E and Pingbing Ming. Cauchy-Born rule and the
            stability of crystalline solids: Static problems. *Arch. Ra-
            tional Mech. Anal.*, 183:241–297, 2007. doi:10.1007/s00205-
            006-0031-7.

[Eri08]     Jerald L. Ericksen.  On the Cauchy-Born rule.  *Math-
            ematics and Mechanics of Solids*, 13(3-4):199–220, 2008.
            doi:10.1177/1081286507086898.

[FL07]      Irene Fonseca and Giovanni Leoni.  *Modern Methods in
            the Calculus of Variations: L^p-Spaces.*  Springer, 2007.
            doi:10.1007/978-0-387-69006-3.

[FS14]      Manuel Friedrich and Bernd Schmidt.  An atomistic-to-
            continuum analysis of crystal cleavage in a two-dimensional
            model problem.  *J. Nonlin. Sci.*,  24:145–183,  2014.
            doi:10.1007/s00332-013-9187-0.

[FT02]      Gero Friesecke and Florian Theil.  Validity and fail-
            ure of the Cauchy-Born hypothesis in a two-dimensional
            mass-spring lattice. *J. Nonlinear Sci.*, 12:445–478, 2002.
            doi:10.1007/s00332-002-0495-z.

[GM12]      Mariano Giaquinta and Luca Martinazzi. *An Introduction to
            the Regularity Theory for Elliptic Systems, Harmonic Maps
            and Minimal Graphs.* Scuola Normale Superiore Pisa, sec-
            onda edizione edition, 2012. doi:10.1007/978-88-7642-443-4.

[GT01]      David Gilbarg and Neil S. Trudinger. *Elliptic Partial Dif-
            ferential Equations of Second Order.* Springer-Verlag Berlin
            Heidelberg, second edition, 2001. doi:10.1007/978-3-642-
            61798-0.

[Hau67]    Siegfried Haussühl. Die Abweichungen von den Cauchy-Relationen. *Phys. kondes. Materie*, 6:181–192, 1967. doi:10.1007/BF02422715.

[HKM77]    Thomas J.R. Hughes, Tosio Kato, and Jerrold E. Marsden. Well-posed quasi-linear second-order hyperbolic systems with applications to nonlinear elastodynamics and general relativity. *Arch. Rational Mech. Anal.*, 63(3):273–294, 1977. doi:10.1007/BF00251584.

[HO12]     Thomas Hudson and Christoph Ortner. On the stability of Bravais lattices and their Cauchy-Born approximations. *ESAIM:M2AN*, 46:81–110, 2012. doi:10.1051/m2an/2011014.

[Mü87]     Stefan Müller. Homogenization of nonconvex integral functionals and cellular elastic materials. *Arch. Rational Mech. Anal.*, 99(3):189–212, 1987. doi:10.1007/BF00284506.

[MPR12]    Nicolas Meunier, Olivier Pantz, and Annie Raoult. Elastic limit of square lattices with three-point interactions. *Math. Models Methods Appl. Sci.*, 22:1250032, 2012. doi:10.1142/S0218202512500327.

[OS12]     Christoph Ortner and Alexander V. Shapeev. Interpolants of lattice functions for the analysis of atomistic/continuum multiscale methods. 2012. URL: http://arxiv.org/abs/1204.3705.

[OT13]     Christoph Ortner and Florian Theil. Justification of the Cauchy-Born approximation of elastodynamics. *Arch. Rational Mech. Anal.*, 207:1025–1073, 2013. doi:10.1007/s00205-012-0592-6.

[Sch08]    Bernd Schmidt. On the passage from atomic to continuum theory for thin films. *Arch. Rational Mech. Anal.*, 190:1–55, 2008. doi:10.1007/s00205-008-0138-0.

[Sch09]    Bernd Schmidt. On the derivation of linear elasticity from atomistic models. *Networks and Heterogeneous Media*, 4(4):789–812, 2009. doi:10.3934/nhm.2009.4.789.

[SS09]     Henry C. Simpson and Scott J. Spector. Applications of estimates near the boundary to regularity of solutions in linearized elasticity. *Siam J. Math. Anal.*, 41(3):923–935, 2009. doi:10.1137/080722990.

[Ste70]    Elias M. Stein. *Singular Integrals and Differentiability Properties of Functions.* Princeton University Press, 1970.

[Str66]    Walter A. Strauss. On continuity of functions with values in various banach spaces. *Pacific J. Math.*, 19(3):543–551, 1966. doi:10.2140/pjm.1966.19.543.

[The11]    Florian Theil. Surface energies in a two-dimensional mass-spring model for crystals. *ESAIM: Mathematical Modelling and Numerical Analysis*, 45:873–899, 2011. doi:10.1051/m2an/2010106.

[Val88]    Tullio Valent. *Boundary Value Problems of Finite Elasticity.* Springer, 1988. doi:10.1007/978-1-4612-3736-5.